T0237222

Towards a Sustainable Future - Life Cycle Management

Zbigniew Stanisław Kłos
Joanna Kałkowska • Jędrzej Kasprzak

Editors

Towards a Sustainable Future - Life Cycle Management

Challenges and Prospects

 Springer

Editors
Zbigniew Stanisław Kłos
Faculty of Civil and Transport Engineering
Poznań University of Technology
Poznań, Poland

Joanna Kałkowska
Faculty of Engineering Management
Poznań University of Technology
Poznań, Poland

Jędrzej Kasprzak
Faculty of Civil and Transport Engineering
Poznań University of Technology
Poznań, Poland

ISBN 978-3-030-77129-4   ISBN 978-3-030-77127-0   (eBook)
https://doi.org/10.1007/978-3-030-77127-0

© The Editor(s) (if applicable) and The Author(s) 2022. This book is an open access publication.

Open Access This book is licensed under the terms of the Creative Commons Attribution 4.0 International License (http://creativecommons.org/licenses/by/4.0/), which permits use, sharing, adaptation, distribution and reproduction in any medium or format, as long as you give appropriate credit to the original author(s) and the source, provide a link to the Creative Commons license and indicate if changes were made.

The images or other third party material in this book are included in the book's Creative Commons license, unless indicated otherwise in a credit line to the material. If material is not included in the book's Creative Commons license and your intended use is not permitted by statutory regulation or exceeds the permitted use, you will need to obtain permission directly from the copyright holder.

The use of general descriptive names, registered names, trademarks, service marks, etc. in this publication does not imply, even in the absence of a specific statement, that such names are exempt from the relevant protective laws and regulations and therefore free for general use.

The publisher, the authors, and the editors are safe to assume that the advice and information in this book are believed to be true and accurate at the date of publication. Neither the publisher nor the authors or the editors give a warranty, expressed or implied, with respect to the material contained herein or for any errors or omissions that may have been made. The publisher remains neutral with regard to jurisdictional claims in published maps and institutional affiliations.

This Springer imprint is published by the registered company Springer Nature Switzerland AG
The registered company address is: Gewerbestrasse 11, 6330 Cham, Switzerland

Preface

Sustainable future – now the dream of poor, middle-income, and rich generations. Therefore, the sustainable development goals are the blueprint for achieving a better and more sustainable future for all. They address the global challenges we face, including environmental degradation and climate change. This book concentrates on issues connected with different aspects influencing environment state and its core idea focused both on the life cycle concept and life cycle management. It includes present contributions on different aspects of making life cycle thinking and product sustainability operational for businesses aiming for continuous improvement, especially striving towards reducing their footprints and minimizing their environmental and socio-economic burdens.

This book presents current scientific and business achievements in the field of life cycle management (LCM), providing methodological and formal development along with practical application, and is the source of theoretical background/knowledge and practical templates. Moreover, some aspects of circular economy–based approaches are highlighted. The book presents new points of view, including regionalities, and it is equipped with the examples coming from around world, especially from all over Europe, including central and eastern European countries.

One of the main ideas of this book is to bring together presentation from the world of science and world of enterprises as well as institutions supporting economic development. This extorts the topics to be more focused on the practical aspects of product life cycle management. The structure of the work is based on five themes. The themes represent different objects and are focused on sustainability and LCM practices mainly related to: products, technologies, organizations, markets, and policy issues as well as methodological solutions.

Undoubtedly, the area of methodological solutions is still of great interest in LCM scientists and practitioners' community. In one chapter, the way in which technology-driven society, facing new environmental challenges, encourages more and more companies' key decision-makers to be committed to limit the impact of their products and services on the environment is presented.

Taking into consideration the fact that ecodesign approaches have shown the potential to increase companies' global value proposition and that the integration of environmental parameters at an early stage of the design process will only be possible if such approach is tailored to a specific sector and customer expectations, the proposal of transposable and replicable action-step methodology facilitating the creation of a common language and enabling the translation of environmental commitments into functional requirements was elaborated and is presented in this work.

At present, an increasing joint use of Life Cycle Assessment and Data Envelopment Analysis (LCA + DEA) as an emerging research field when evaluating many similar entities in the framework of eco-efficiency and sustainability, may be observed. In addition, an interesting attempt to enhance life cycle management through the symbiotic use of data envelopment analysis, showing innovative advances in LCA + DEA analyses, is presented. It reveals the situation within the tertiary sector, exploring the novel advances offered regarding the application of the well-established five-step method for enhanced sustainability benchmarking.

The main interests for LCM practitioners are focused on products area – an interesting example is presented in the chapter showing life cycle assessment benchmark for wooden buildings in Europe. Despite the fact that LCA and the EU-recommended environmental footprints (EF) are well known and accepted tools to measure a comprehensive set of environmental impacts throughout a product's life cycle, the assessment of level of environmental performance of wooden buildings is still a challenge. Based upon the EU recommendations for a benchmark of all kinds of European dwellings, a scenario of a typical European wooden building was developed. The developed benchmark for wooden buildings is a suitable comparison point for new wooden building designs. This benchmark can be used by architects and designers early in the planning stages – when changes can still be made to improve the environmental performance or to communicate and interpret LCA results for customers and other stakeholders.

As usual, there are several chapters focused on sustainability and LCM practices related to different technologies. Some of them concentrated on different aspects of analyses dealing with municipal solid waste management. Among others there is the presentation of framework for the systematic analysis of the material flows and the life cycle environmental performance of municipal solid waste management scenarios. This framework is capable of predicting the response of waste treatment processes to the changes in waste streams composition that inevitably arise in municipal solid waste management systems. The fundamental idea is that the inputs and outputs into or from treatment processes are previously allocated to the specific waste materials contained in the input waste stream. Aggregated indicators like life cycle environmental impacts can then be allocated to waste materials, allowing systematic scenario analyses. The given chapter framework is generic and flexible, and can easily be adapted to other types of assessments such as economic analysis and optimization.

Specific aspects of issues focused on sustainability and LCM practices related to application in different organizations are also presented in several chapters. In one case study, an interesting dilemma, "is environmental efficiency compatible with economic competitiveness in dairy farming?" is considered on the base of study of Luxembourgish farms. The main aim of this study is to investigate both environmental and economic performances of dairy farms in order to assess possibilities and limits of improving economic competitiveness via increasing environmental efficiency. Analysis of four LCA-impact environmental categories and three economic ones shows that a sustainable dairy production with less environmental impact in all considered categories is also of advantage in terms of farm economic competitiveness. The case study proves that a high environmental performance is not only of advantage in terms of economic competitiveness, but is even a necessary prerequisite for best economic performances. Other interesting application of LCM in organization is shown on the example of international non-governmental organization practicing in the social and environmental sector. Thanks to environmental focus considered NGO is interested in assessing the environmental impacts of its own activities throughout the whole value chain and therefore, an Organizational Life Cycle Assessment study had been conducted for one NGO community.

Markets and policy issues are specific field for LCM application. Among others, specific problems generate the fact that wind power generation is weather-dependent and that at a high penetration rate, storage systems such as power-to-gas may become necessary to adjust electricity production to consumption. One chapter presents the environmental life-cycle performance of wind power accounting for the energy storage induced by the temporal variability of weather-dependent production and consumption. A case study in which wind power installations are combined with a power-to-gas system in Denmark to provide electricity according to the national load consumption profile is analyzed. Results highlight an increase, roughly by a factor two, of the carbon footprint coming from both energy storage infrastructure and induced losses, but remain significantly, at least ten times, lower than fossil counterparts.

The content of this monograph is the presentation of various examples of scientific and practical contributions, showing the incorporation of a life cycle approach into decision-making processes at the strategic and operational level. Special attention is drawn to show how to apply LCM to target, organize, analyze, and manage product-related information and activities towards continuous improvement, along the different products life cycle. The panorama of cases presents that LCM is a business management approach that can be used by all types of businesses and organizations in order to improve their sustainability performance. Thus, this book provides a cross-sectoral, present picture in LCM issues area.

The book presents chapters by scientists and researchers working in the field of sustainable development who have engaged in dynamic approaches to implementing widely understood sustainability. It guides the reader to understand the current issues of life cycle management and can be used apply product-related information to ensure more sustainable value chain management. The book encompasses many practical, methodological, and theoretical fields. Recently introduced to the technical and business vocabulary, term "LCM" is about a business management approach that can be used by all types of business and other organizations in order to improve their products' sustainability performance to promote better understanding among students and business professionals in the utility sector and across industries.

Poznań, Poland

Zbigniew Stanisław Kłos

Joanna Kałkowska

Jędrzej Kasprzak

Acknowledgments

This book is a selection of contributions of LCM 2019 conference held in Poznań, Poland. The series of life cycle management conferences is established as one of the leading events worldwide in the area of environmental, economic, and social sustainability. The 2019 conference motto "Towards sustainable future – current challenges and prospects in life cycle management" was chosen to focus on both current challenges and current prospects in the LCM field. In total, 502 abstracts from 41 countries were submitted. High quality of the contributions made it quite a challenge to select 26 papers for this book. They have been organized into five parts.

Due to core contribution of the authors and, in addition, due to the efforts of many colleagues and friends, this book became possible. We are very grateful for the support by members of scientific and business committees, namely acting as a co-chairs of this body: Anna Lewandowska, Joanna Kulczycka, Stefan Trzcieliński, and Zbigniew Kłos, as well as: Can B. Aktas, Stefan Albrecht, Joana Almeida, Martin Baitz, Fritz Balkau, Ligia Tiruta-Barna, Christian Bauer, Sandra Belboom, Catherine Benoit Norris, Alberto Bezama, Jan Bollen, Miguel Brandao, Birgit Brunklaus, Dorota Burchart-Korol, Julie Clavreul, Pierre Collet, Mauro Cordella, Helena Dahlbo, Francesco Del Pero, Andrea Di Maria, Johannes Drielsma, Jim Fava, Matthias Finkbeiner, Göran Finnveden, Matthias Fischer, Zenon Foltynowicz, Michele Galatola, Hans Garvens, Mark Goedkoop, Jeroen Guinée, Karen Hanghøj, Krzysztof Hankiewicz, Aranud Hélias, Almudena Hospido, Christine Roxanne Hung, Diego Iribarren, René Itten, Allan Astrup Jensen, Joanna Kałkowska, Renata Kaps, Kamil Kasner, Robert Kasner, Jędrzej Kasprzak, Regula Keller, Nicola Kimm, Martin Kirchner, Stephan Krinke, Michał Kulak, Martin Kurdve, Karin Lochte, Tim Lohse, Søren Løkke, Antonino Marvuglia, Paulo Masoni, Eric Mieras, Mark Mistry, Tiago G. Morais, Dominik Muller, Magdalena Muradin, Belmira Neto, Hanna Nilsson-Lindén, Mikołaj Owsianiak, Tiina Pajula, Elisabetta Palumbo, Nicholas Perry, Anna Petit-Boix, Marina Proske, Francesca Recanati, Patricia Reis, Emma Rex, Marcin Sadowski, Peter Saling, Roberta Salomone, Liselotte Schebek, Andreas Schiffleitner, Karsten Schischke, Sergiy Smetana, Guido Sonnemann, Dariusz Stasik, Sangwon Suh, Wojciech Sumelka, Nydia Suppen-Reynaga, Marzia Traverso, Georgios Tsimiklis, Sonia Valdivia, Marisa Vieira, Andy Whiting, Pia

Wiche, Joanna Witczak, Joanna Zarebska, Krzysztof Zarnotal, Rosalie Van Zelm, and Stefano Zuin. Their efforts in selecting the right mix of contributions were strongly valuable. Unfortunately the coronavirus disease influenced on the rhythm of preparation of this book, significantly prolonging this process.

We also want to thank two research group members at Poznań University of Technology and one research group member at Poznań University of Economics for all their support. From the very beginning, Krzysztof Koper was deeply engaged in organization of the 9th International Conference; he unfortunately left us before the conference. We also would like to acknowledge the technical support of Michał Trziszka, Krzysztof Hankiewicz, and Grzegorz Musioł.

Lastly, we direct our sincere thanks to our families for their patience and courtesy.

Zbigniew Stanisław Kłos

Joanna Kałkowska

Jędrzej Kasprzak

Contents

Contributors

Boris Agarski Faculty of Technical Sciences, University of Novi Sad, Novi Sad, Serbia

Franck Aggeri CGS-i3 Mines-ParisTech, PSL University, Paris, France

Can B. Aktaş Department of Civil Engineering, TED University, Ankara, Turkey

Cristina Álvarez-Rodríguez Rey Juan Carlos University, Chemical and Environmental Engineering Group, Móstoles, Spain

Patrycja Baldowska-Witos Department of Technical Systems Engineering, Faculty of Mechanical Engineering, University of Science and Technology in Bydgoszcz, Bydgoszcz, Poland

Caroline Battheu-Noirfalise Centre Wallon de Recherches Agronomiques (CRA-W), Libramont, Belgium

Romain Besseau Mines ParisTech, PSL University, Paris, France

Bogusław Bieda Department of Management, AGH University of Science and Technology, Kraków, Poland

Isabelle Blanc Mines ParisTech, PSL University, Paris, France

Katrien Boonen VITO NV, Sustainable Materials Management Unit, Mol, Belgium

Igor Budak Faculty of Technical Sciences, University of Novi Sad, Novi Sad, Serbia

Juan Pablo Chargoy-Amador Center for Life Cycle Assessment and Sustainable Design (CADIS), Life Cycle Management Director, Mexico City, Mexico

Edivan Cherubini EnCiclo Soluções Sustentáveis Ltda., Florianópolis, Brazil

Jorge Delgado Braskem S.A., São Paulo, Brazil

Milien Dhorne Mines ParisTech, PSL University, Paris, France

Ricardo Dias Braskem S.A., São Paulo, Brazil

Nicholas Dodd European Commission, Joint Research Centre, Seville, Spain

Javier Dufour IMDEA Energy, Systems Analysis Unit, Móstoles, Spain
Rey Juan Carlos University, Chemical and Environmental Engineering Group,
Móstoles, Spain

Tom Dusseldorf CONVIS Société Coopérative, Ettelbruck, Luxembourg

Nieves Espinosa European Commission, Joint Research Centre, Seville, Spain

Horst Fehrenbach ifeu – Institute for Energy and Environmental Research
Heidelberg, Heidelberg, Germany

Zenon Foltynowicz Poznań University of Economics and Business, Poznań, Poland

José-Luis Gálvez-Martos IMDEA Energy, Systems Analysis Unit,
Móstoles, Spain

Tatiana Abaurre Alencar Gavric InnoRenew Centre of Excellence (CoE), Izola -
Isola, Slovenia

José Manuel Gil-Valle LCI Member (Private advisor), Etagnières, Switzerland

Sebastian Glaser Technische Universität Wien, Vienna, Austria

Jörn Hartung Siemens Corporate Technology, Research in Energy and
Electronics, Energy Systems, Sustainable Life Cycle Management and
Environmental Performance Management, Berlin, Germany

Klaus Heidinger Siemens AG Austria, Siemensstrasse, Vienna, Austria

Marie Hemmen ifeu – Institute for Energy and Environmental Research
Heidelberg, Heidelberg, Germany

Jens-Christian Holst Siemens AG, Corporate Technology, Berlin, Germany

Nabila Iken CGS-i3 Mines-ParisTech, PSL University, Paris, France

Diego Iribarren IMDEA Energy, Systems Analysis Unit, Móstoles, Spain

Ioan-Robert Istrate IMDEA Energy, Systems Analysis Unit, Móstoles, Spain
Rey Juan Carlos University, Chemical and Environmental Engineering Group,
Móstoles, Spain

Norihiro Itsubo Department of Environmental and Information Studies, TCU
Tokyo City University, Yokohama, Japan

Florian Ansgar Jaeger Siemens Corporate Technology, Research in Energy and
Electronics, Energy Systems, Sustainable Life Cycle Management and
Environmental Performance Management, Berlin, Germany

Mats Johansson Department of Technology, Management and Economics, Chalmers University of Technology, Göteborg, Sweden

Yuki Kabe Braskem S.A., São Paulo, Brazil

Robert Kasner Department of Technical Systems Engineering, Faculty of Mechanical Engineering, University of Science and Technology in Bydgoszcz, Bydgoszcz, Poland

Heiko Keller ifeu – Institute for Energy and Environmental Research Heidelberg, Heidelberg, Germany

Zbigniew Kłos Poznań University of Technology, Poznań, Poland

Wilhelm Kuckshinrichs Forschungszentrum Jülich, Institute of Energy and Climate Research – Systems Analysis and Technology Evaluation, Jülich, Germany

Yoko Kurahara Department of Environmental and Information Studies, TCU Tokyo City University, Yokohama, Japan

Martin Kurdve RISE IVF – Research Institutes of Sweden, Stockholm, Sweden Department of Technology, Management and Economics, Chalmers University of Technology, Göteborg, Sweden

Kévin Le Blévennec VITO NV, Sustainable Materials Management Unit, Mol, Belgium

Aline Lehnen CONVIS Société Coopérative, Ettelbruck, Luxembourg

Rocco Lioy CONVIS Société Coopérative, Ettelbruck, Luxembourg

Mario Martín-Gamboa Department of Environment and Planning, University of Aveiro, Centre for Environmental and Marine Studies (CESAM), Aveiro, Portugal

Milana Ilic Micunovic Faculty of Technical Sciences, University of Novi Sad, Novi Sad, Serbia

Stéphane Morel Alliance Technology Development, Renault sas, Guyancourt, France

Wladmir H. Motta CEFET-RJ, Rio de Janeiro, Brazil

Katrin Müller Siemens Corporate Technology, Research in Energy and Electronics, Energy Systems, Sustainable Life Cycle Management and Environmental Performance Management, Berlin, Germany

Magdalena Muradin Poznan University of Economics and Business, Poznan, Poland

Sabrina Neugebauer RWTH Aachen University, Aachen, Germany

Aarne Johannes Niemelä InnoRenew Centre of Excellence (CoE), Izola - Isola, Slovenia

Eva Prelovšek Niemelä InnoRenew Centre of Excellence (CoE), Izola - Isola, Slovenia

Nils F. Nissen Fraunhofer Institute for Reliability and Microintegration (IZM), Berlin, Germany

Elisabetta Palumbo RWTH Aachen University, Institute of Sustainability in Civil Engineering (INaB), Aachen, Germany

Rainer Pamminger Technische Universität Wien, Vienna, Austria

Paula Pérez-López Mines ParisTech, PSL University, Paris, France

Marina Proske Fraunhofer Institute for Reliability and Microintegration (IZM), Berlin, Germany
Technische Universität Berlin, Research Center Microperipheric Technologies, Berlin, Germany

Roman Reding CONVIS Société Coopérative, Ettelbruck, Luxembourg

Nils Rettenmaier ifeu – Institute for Energy and Environmental Research Heidelberg, Heidelberg, Germany

Dariusz Sala Department of Management, AGH University of Science and Technology, Kraków, Poland

Erwin M. Schau InnoRenew Centre of Excellence (CoE), Izola - Isola, Slovenia

Karsten Schischke Fraunhofer Institute for Reliability and Microintegration (IZM), Berlin, Germany

Martin Schneider-Ramelow Fraunhofer Institute for Reliability and Microintegration (IZM), Berlin, Germany
Technische Universität Berlin, Berlin, Germany

Andrea Schreiber Forschungszentrum Jülich, Institute of Energy and Climate Research – Systems Analysis and Technology Evaluation, Jülich, Germany

Sasha Shahbazi RISE IVF – Research Institutes of Sweden, Stockholm, Sweden

Cornelia Sonntag Siemens Corporate Technology, Research in Energy and Electronics, Energy Systems, Sustainable Life Cycle Management and Environmental Performance Management, Berlin, Germany

Iztok Šušteršič InnoRenew Centre of Excellence (CoE), Izola - Isola, Slovenia

Andrzej Tomporowski Department of Technical Systems Engineering, Faculty of Mechanical Engineering, University of Science and Technology in Bydgoszcz, Bydgoszcz, Poland

Maria Chiara Torricelli University of Florence, Florence, Italy

Marzia Traverso RWTH Aachen University, Institute of Sustainability in Civil Engineering (INaB), Aachen, Germany

Dejan Ubavin Faculty of Technical Sciences, University of Novi Sad, Novi Sad, Serbia

Patricia van Loon Chalmers Industriteknik, Göteborg, Sweden

Mabel Vega-Coloma Escuela de Ingeniería Química, Facultad de Ingeniería, Universidad del Bío-Bío, Concepción, Chile

An Vercalsteren VITO NV, Sustainable Materials Management Unit, Mol, Belgium

Alejandro Villanueva European Commission, Joint Research Centre, Seville, Spain

Djordje Vukelic Faculty of Technical Sciences, University of Novi Sad, Novi Sad, Serbia

Christina Wulf Forschungszentrum Jülich, Institute of Energy and Climate Research – Systems Analysis and Technology Evaluation, Jülich, Germany

Guilherme Zanghelini EnCiclo Soluções Sustentáveis Ltda., Florianópolis, Brazil

Petra Zapp Forschungszentrum Jülich, Institute of Energy and Climate Research – Systems Analysis and Technology Evaluation, Jülich, Germany

Claudio Zaror Departamento de Ingeniería Química, Facultad de Ingeniería, Universidad de Concepción, Concepción, Chile

List of Figures

Role of Stochastic Approach Applied to Life Cycle Inventory (LCI) of Rare Earth Elements (REEs) from Secondary Sources Case Studies

Extending LCA Methodology for Assessing Liquid Biofuels by Phosphate Resource Depletion and Attributional Land Use/Land Use Change

List of Tables

Part I
Sustainable Products

Ecodesign as a New Lever to Enhance the Global Value Proposition: From Space to Corporate

Kévin Le Blévennec, An Vercalsteren, and Katrien Boonen

Abstract As our technology-driven society is facing new environmental challenges, more and more companies' key decision-makers are committing to limit the impact of their products and services on the environment. While ecodesign approaches have shown the potential to increase companies' global value proposition, the integration of environmental parameters at an early stage of the design process will only be possible if such approach is tailored to a specific sector and customer expectations. To support environmental experts in charge of organizing the integration of such approach in the design process of complex engineering systems, VITO retrospectively analysed a project initiated by the European Space Agency (ESA), from a product strategy perspective. A transposable and replicable action-step methodology facilitating the creation of a common language and enabling the translation of environmental commitments into functional requirements is resulting.

1 Introduction

1.1 Principles of Responsibility Driving Stronger Environmental Commitments

Launched in 2000, the United Nations Global Compact (UNGC) is both a policy platform and a practical framework for companies that are committed to sustainable business practices. It seeks to align business operations and strategies everywhere with ten universally accepted principles in support of achieving the Sustainable Development Goals by 2030. Over 12,000 organizations around the world, including 9953 companies, have joined the UNGC. Principles number 8, 9 and 10, respectively, mention that businesses should support a precautionary approach to environmental challenges, undertake initiatives to promote greater environmental

K. Le Blévennec (✉) · A. Vercalsteren · K. Boonen
VITO NV, Sustainable Materials Management Unit, Mol, Belgium
e-mail: kevin.leblevennec@vito.be

© The Author(s) 2022
Z. S. Klos et al. (eds.), *Towards a Sustainable Future - Life Cycle Management*,
https://doi.org/10.1007/978-3-030-77127-0_1

3

responsibility and encourage the development and diffusion of environmentally friendly technologies [1].

Within this global context, the objectives set in the EU's seventh Environmental Action Programme, entitled 'Living well, within the limits of our planet', gave rise to a political focus on the circular economy. The resulting Circular Economy Action Plan includes several measures covering the whole material cycle, from production and consumption to waste management and the market for secondary raw materials. The proposed actions, highlighting the essential role of the private sector, aim to contribute to the transition to a circular economy by 'closing the loop' of product life cycles through an increase of recycling and reuse, benefiting both the environment and the economy [2].

By signing the UNGC, companies are committing to a deliberate, responsible approach to the protection of the environment. It is often reflected in a policy to reduce environmental impacts and risks in various activities of the company. Due to the environmental challenges that our society is facing, new stakeholders such as investors are getting concerned and involved. Not only to comply with this evolving political framework, many key decision-makers have considered their commitment to the UNGC as a driver of ambitious environmental policies shaping new opportunities in their business practices and enhancing their global value proposition. To reinforce their commitment, goals are set in terms of energy, climate, waste, environmental management of the supply chain and increasingly concerning product design. More and more companies are now committing to a responsible approach which aims at limiting the impact of their products and services on the environment.

The role of product design is a key in the transition to a circular economy; the effective implementation of those environmental policies now targeting products could thus accelerate this transition. By recirculating products instead of throwing them out, not only is the value of products and components retained, but also the demand for virgin materials decreases, as do the energy demand and the production of (hazardous) waste. Product design heavily influences a product's life cycle impacts and is crucial for connecting different stages and actors along the life cycle.

1.2 From a Commitment to an Effective Implementation?

In many sectors, the regulatory and normative contexts as well as the pressure of markets are leading more and more companies to deepen a process characterized by the integration of environmental parameters since the design of products. Since the 1990s, engineers and designers, in joint efforts and gradually enriched, have defined an approach to those environmental concerns: ecodesign also defined as design for environment but with the same meaning – 'design with environment' [3].

In many situations, ecodesign demonstrated its effectiveness. By implementing this activity, companies have managed to reduce their costs, access new markets, develop new partnerships and arouse new investors' interest. Not only reducing products environmental impact, it has been proved as a concrete source of industrial competitiveness [4].

While benefits might be multidimensional, many ecodesign approaches also showed some limits. Their integration within existing company practices is key. Only a few organizational tools exist and are with difficulty transposable from one company to another. Integration of environmental parameters needs to be done in a collaborative way and using suitable tools, chosen in function of the company's size, maturity regarding ecodesign and existing design processes. The one off intervention as well as the development of tools and methods only intelligible by environmental experts has sometimes showed the limit of ecodesign integration within companies. When implementing ecodesign activities, a deep understanding of the current business context has also been listed as a key success factor [5–7].

1.3 Challenges Related to Constrained Engineering Environments

Many authors have emphasized the importance of the business context: the integration of an ecodesign approach is progressive and needs to be tailored to the company's sector and customer expectations. If effective ecodesign approaches have shown their effectiveness in mass-consumer goods producers, the situation is not so advanced in business-to-business industries, even less for companies developing complex engineering systems.

In sectors such as the space and aerospace sectors, many companies are developing and offering a range of solutions along the critical decision chain. Their activities are mainly referring to business-to-business and/or business-to-government with customers oriented towards security, reliability and performance. Those solutions are evolving in a complex environment requiring an integration of an important number of functionalities and constraints to face up to any events and ensuring customers' security. Due to a possible evolving customer demand during the design process or again due to high prototypes costs, concurrent engineering methodologies are applied to ensure the critical dependencies of defined functional requirements. Complexity of those solutions mainly implies that those organizations are characterized by multiple highly specialized experts who infrequently communicate together [8].

Driven by performance and technological considerations, those companies are part of highly competitive markets in which value creation is at the heart of the concerns. The previous limited number of customer requests with regard to ecodesign and the specificity of product development life cycles did not lead most of internal actors to collaborate and anticipate this environmental parameter integration into design processes. While ecodesign approaches have shown the potential to increase companies' global value proposition, the integration of environmental parameters in the design process can still be considered as a constraint by many key actors. The following research question is thus appearing: how to organize the integration of an efficient ecodesign into the current design process of complex engineering systems?

2 Retrospective Analysis of the GreenSat Project

2.1 Objective and Methodological Approach

In order to reduce the risk of inactions of companies evolving in constrained engineering environment which have committed to limit the impact of their products and services on the environment, or again limit divergent or unfinished actions within those companies, the objective of this research is to develop a methodology supporting environmental experts in charge of organizing the integration of an ecodesign approach into the current design process of the solutions developed by their company.

Being the archetype of a constrained engineering environment, a case study within the space sector has been selected. In the GreenSat project initiated by the European Space Agency (ESA), VITO was commissioned (together with QinetiQ) to identify and select ecodesign options for the PROBA-V mission. A brief introduction to this study, focusing on the different steps having enabled the achievement of the project rather than on the results themselves, is provided in Sect. 2.2. Those results were fully described in the final report of the study [9].

Based on this case study approach and within the framework of its 'design for the circular economy' strategic research activities, VITO has retrospectively analysed this project from a product strategy perspective. The learnings and outcomes are reported in Sect. 2.3.

2.2 Introduction to the GreenSat Project

PROBA (PRoject for On-Board Autonomy) is a family of small satellites developed for ESA by QinetiQ Space and launched in 2013. The PROBA-Vegetation (PROBA-V) mission, an earth observation mission, was selected as a continuation of the Vegetation programme. The main payload of the PROBA-V satellite is the Vegetation instrument. Through the GreenSat project, ESA wanted to evolve from assessment to reduction of environmental impact through the redesign of an existing satellite mission and to check the feasibility of implementing ecodesign in the development of future space missions.

In a first step of the GreenSat project, a life cycle assessment (LCA) has been performed with the following objectives:

1. To identify environmental hotspots of the mission, which was considered as an important starting point to look for ecodesign options.
2. To quantify the environmental impact of the mission, to understand the impacts and the sources, which is a baseline to benchmark the environmental impact of the ecodesigned GreenSat mission and which allows to assess the environmental impact reduction.

Concerning the redesign of the mission, ESA identified initial requirements that would need to be adapted in case of a 'GreenSat' PROBA-V mission. In particular, the system should achieve equivalent function, meaning that the functional requirements should be almost all the same. As a consequence, while functional requirements should not be significantly different, design and operations requirements however could be significantly different.

The functional unit has been defined conforming to the space system LCA guidelines: 'one space mission in fulfilment of the mission's requirements'. The PROBA-V LCA includes all activities in the space and ground segment, but the launch segment was excluded. The life cycle impact assessment (LCIA) for PROBA-V was performed for the environmental impact categories and according to the defined LCIA methods as provided in the LCIA method in the ESA LCA database [10]. Given the relevance of critical materials use in space applications, an additional 'impact category (criticality – weighted)' that assesses the availability of raw materials, taking into account socio-economic constraints, has been defined.

Environmental profile of the PROBA-V mission was thus obtained. The LCA allowed identifying the environmental hotspots for the PROBA-V mission, which were considered most relevant to look at for the ecodesign exercise in the next study phase. Hotspots per mission phase were identified, and a distinction was made between the different levels the hotspot can relate to: materials, equipment and components, manufacturing processes, system, management and programmatic issues and regulation. If an environmental hotspot contributes to more than one impact category, its environmental importance was considered higher (Fig. 1).

As ESA has only little influence in the ground segment activities (e.g. energy use, data processing equipment), it was considered essential, and it has been decided to focus the ecodesign phase of the study on technologies where ESA has impact on. Starting from the identified environmental hotspots, ecodesign options for improving the environmental performance have been defined. For that, an external workshop was first organized at ESA concurrent design facility premises, with a wider group of stakeholders. Then an internal brainstorm was coordinated at QinetiQ Space, with experts specifically involved in the PROBA-V life cycle stages. A long list of ecodesign options was generated for space missions in general and PROBA-V in particular. As only a limited number of ecodesign options could be further developed in the framework of the GreenSat project, a selection process was applied to the full list of options.

As a first step, the analytic hierarchy process (AHP) methodology was used to select the 25 most promising options among the 70+ identified. AHP allows calculating weighting factors for the trade-off criteria, based on input from different

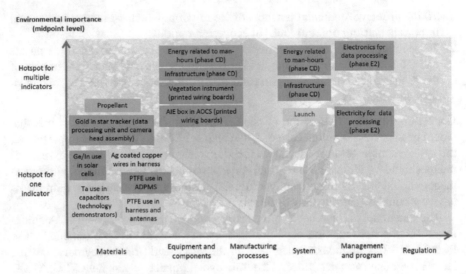

Fig. 1 Hotspots of the PROBA-V mission

stakeholders (brainstorm participants). Scores are then assigned to each ecodesign option by VITO and QinetiQ, which leads to a final ranking of all ecodesign options. This trade-off was based on specific criteria:

1. Solution implementation effort (cost, manhours, means)
2. Duration (time to market/launch)
3. Risk (feasibility, applicability, performance, availability of alternatives, flexibility)
4. Impact (operational cost)
5. Overall environmental impact
6. Reusability of the solution
7. Space relevance (technologies for the space segment, activities related to the space segment and their preparation that differ from other ground activities, system approach)

In a next step, a semi-quantitative analysis was done to assess the potential reduction of each option, leading to the down-selection of ten options. In a third step, an estimate of the required design and development effort and related environmental impact to estimate the risk of burden shifting was initiated, leading to the final selection of six options.

A technical assessment was then performed, and an analysis of the environmental effects was done by LCA. The analysis was done per option, at different levels (e.g. material, satellite production and mission). Finally, a combined analysis of the environmental impacts including the three most promising options, improved data processing, more sustainably produced germanium and optimized electronics, was performed to assess the overall reduction potential of the redesign of the PROBA-V mission.

Overall, it was concluded that the implementation of ecodesign from the start of a space mission design and development process can actually reduce the environmental impact of the space mission significantly. It was recommended to focus efforts in a first instance on the environmental hotspots of a space mission as this leads to the largest improvements. Improvements are related not only to the satellite production but also to the operational phase of the satellite (e.g. data processing). In addition to a final report, the results have been summarized and communicated through an infographic.

2.3 Retrospective Analysis from a Product Strategy Perspective

VITO aims to accelerate the transition towards a circular economy by providing advice to companies. As product design is key for enabling circularity, VITO decided to capitalize on the achievement of the GreenSat project for developing a methodology supporting environmental experts in charge of organizing the integration of an ecodesign approach into the current design process of complex engineering systems. An internal brainstorm session between 'circular economy assessment methods and indicators' and 'design for the circular economy' VITO experts has been organized. The objective of this session was to bring face to face those two research perspectives in order to detect essential factors for the development of the methodology.

Outcomes of this brainstorm session have been summarized (Tables 1 and 2). Elements listed in the project perspective (Table 1) refer to a specific action undertaken during the GreenSat project, as described in Sect. 2.2. Corresponding elements in the product strategy perspective (Table 2) refer to a derived analogy between this specific action and identified key success factors for an effective implementation of an ecodesign approach, as introduced in Sects. 1.2 and 1.3.

3 A Transposable and Replicable Resulting Methodology

The retrospective analysis of the GreenSat project from a product strategy perspective allowed illustrating key success factors for efficiently implementing an ecodesign approach with concrete specific actions undertaken during a project within a constrained engineering environment. Based on those analogies (Tables 1 and 2), the creation of a common language ensuring the compatibility of the stakeholders' different sources of value creation appears to be essential for efficiently implementing an ecodesign approach. Necessary actions facilitating the creation of this common language thus originate from this analysis.

By organizing those necessary actions and generalizing some elements specific to the GreenSat project, the following action-step methodology enabling the creation of this common language has been developed (Table 3).

Table 1 Outcomes of internal brainstorm session bringing face to face two research perspectives – part 1

Element	Project perspective
1.	[…] project initiated by the European Space Agency (ESA), VITO was commissioned by QinetiQ […]
2.	[…] With the GreenSat project, ESA wanted to check the feasibility of implementing ecodesign in the development of future space missions […]
3.	[…] The system should achieve equivalent function, meaning that the functional requirements should be almost all the same […]
4.	[…] Conform to the space system LCA guidelines […], […] ESA LCA database […], […] Given the relevance of critical materials use in space applications, an additional 'impact category' […] has been defined […]
5.	Figure 1 Hotspots of the PROBA-V Mission, […] communicated through an infographic […]
6.	[…] As ESA has only little influence in the ground segment activities (e.g. energy use, data processing equipment), it was considered essential, and it has been decided to focus the ecodesign phase of the study on technologies where ESA has impact on […]
7.	[…] An external workshop was first organized at ESA concurrent design facility premises, with a wider group of stakeholders. Then an internal brainstorm was coordinated at QinetiQ Space, with experts specifically involved in the PROBA-V life cycle stages […]
8.	Analytic Hierarchy Process (AHP) methodology
9.	Specific criteria used for the AHP methodology
10.	[…] Per option, a technical assessment was performed and an analysis of the environmental effects was done by LCA […]
11.	[…] Implementation of ecodesign from the start of a space mission design and development process can actually reduce the environmental impact of the space mission significantly […]

All these actions-steps are further explained:

1. The project was initiated by ESA which wanted to reduce the environmental impact of their product. ESA through the ESA Clean Space initiative supports the UNGC, having a Committee on the Peaceful Uses of Outer Space: 'States and international intergovernmental organizations should promote the development of technologies that minimize the environmental impact of manufacturing and launching space assets and that maximize the use of renewable resources and the reusability or repurposing of space assets to enhance the long-term sustainability of those activities' [11]. The commitment of key decision-makers is thus the real starting point of this project. To integrate environmental parameters at an early stage of a design process, it is essential to have the internal support of companies' key decision-makers. To convince them, it will be essential to demonstrate that the ecodesign approach will benefit the company's environmental strategy. Listing all environmental commitments associated with products and services is thus considered as a first step.

2. The GreenSat project reflected an efficient external collaboration between three main stakeholders. Based on identified environmental commitments associated

Table 2 Outcomes of internal brainstorm session bringing face to face two research perspectives – part 2

Element	Product strategy perspective
1.	External collaboration between three stakeholders. ESA being the customer. QinetiQ being the industrial partner collaborating with VITO for their environmental expertise
2.	Description of the project objective showing a customer interested in the implementation of an ecodesign approach
3.	System driven by performance. While having the objective to reduce the environmental impacts, environmental and industrial sources of value creation should be compatible
4.	Tailoring of existing tools (guidelines, database, impact category) to a specific business context, i.e. space sector
5.	Intelligibility of results adapted to interested stakeholders
6.	Focus on customer sources of value creation
7.	Collaboration between highly specialized experts from different fields
8.	Use of a specific methodology for connecting highly specialized experts' inputs
9.	Criteria including parameters being source of value creation for all stakeholders: Customer, industrial, environmental
10.	Ensure the compatibility of industrial and environmental sources of value creation
11.	Early stage integration of environmental parameters

Table 3 Definition of an action-step methodology for organizing the integration of an efficient ecodesign approach

Number	Action step
1	List environmental commitments associated with products and services
2	Identify the right stakeholders
3	Understand individual value creation processes
4	Connect individual value creation processes
5	Deliver efficient messages

with products and services, it is essential to understand what expertise(s) and thus which stakeholder(s) will be required to collaborate for achieving those commitments. For instance, if a commitment refers to the use of a minimum recycled content for materials of specific products, involving the engineering teams, the marketing and/or the purchasing department might be relevant. If relevant expertise cannot be identified internally, involving the right external stakeholders is also determining.

3. The retrospective analysis highlighted that a factor for ensuring an external collaboration between highly specialized experts from different fields, with regard to the integration of environmental parameters, was to ensure the compatibility between different sources of value creation. In this case study, the search of ecodesign options was, for instance, reduced to ESA's scope of actions, and the AHP methodology was integrating parameters' source of industrial competitiveness but also benefiting the environment. Before ensuring their compatibility, it

is thus essential to understand individual 'business languages' and how each of the identified stakeholders individually creates value for their company or own departments within an organization. An analysis of individual stakeholders' value creation processes is thus recommended.

4. Once individual stakeholders' value creation processes are understood, the objective is to ensure their compatibility. For that, there is a need to develop tailored tools and practices ensuring the creation of a common language between those stakeholders. ESA LCA guidelines and database are a concrete example. The space sector has unique characteristics like low production rates, long development cycles and specialized materials and processes, and the sector's activities create impacts on environments generally not considered in LCA. Issued in 2016, these guidelines aim to establish methodological rules for performing space-specific LCA. A methodology mostly used by environmental experts is thus tailored to a specific business context thus facilitating the communication and exchange of information with highly specialized engineers.

5. Also shown with the analysis of the GreenSat project, intelligibility of results was considered as determining in this collaborative process. The described matrix (Fig. 1) is, for instance, translating results often only intelligible by LCA experts into clear messages in a language perfectly intelligible for a system engineer. Once tailored tools and practices have been developed, it is thus key to ensure their effective use. The last step of this methodology thus emphasizes the fact that results need to be illustrated and intelligible for non-environmental experts. This last step is also a key for delivering convincing arguments to relevant stakeholders.

By following the different steps of this methodology, the translation of environmental commitments into functional requirements should be facilitated for environmental experts in charge of organizing the implementation of an efficient ecodesign approach into their companies.

4 Conclusion

This case study within the space sector demonstrated that implementing an ecodesign approach could enhance companies' global value proposition by ensuring a compatibility of stakeholders' sources of value creation during the integration of environmental parameters at an early stage of the design process. Originating from a retrospective analysis from a product strategy perspective of this space project, an action-step methodology supporting environmental experts in charge of organizing the integration of an ecodesign approach into the current design process of the solutions developed by their company has been defined. While arising from a case study in a constrained engineering environment, this methodology could be replicated in different sectors and applicable to external as well as internal collaborations. In a next applied research step, 'design for the circular economy' VITO experts aim to test and validate this methodology with different use cases to finally propose this methodology as a service, in order to support the private sector in accelerating the transition towards a more circular economy.

References

1. https://www.unglobalcompact.org/what-is-gc/mission/principles. Accessed 14 Jan 2020.
2. European Commission. (2015). *Closing the loop – An EU action plan for the circular economy*. European Commission.
3. Abrassart, C., & Aggeri, F. (2002). La naissance de l'éco-conception. Du cycle de vie au management environnemental produit. *Responsabilité et Environnement – Annales des Mines, 25*, 14–63.
4. Charter, M. (1997). Managing eco-design. *UNEP Industry and Environment, 20*, 29–31.
5. Knight, P., & Jenkins, J. (2009). Adopting and applying eco-design techniques: A practitioners perspective. *Journal of Cleaner Production, 17*(5), 549–558.
6. Pascual, O., Boks, C., & Stevels, A. (2003). Communicating eco-efficiency in industrial contexts: A framework for understanding the (lack) of success and applicability of eco-design. In *IEEE international symposium on electronics and the environment* (pp. 303–308). IEEE Computer Society.
7. Domingo, L., Buckingham, M., Dekoninck, E., & Cornwell, H. (2015). The importance of understanding the business context when planning eco-design activities. *Journal of Industrial and Production Engineering, 32*(1), 3–11.
8. Cluzel, F., Yannou, B., Leroy, Y., & Millet, D. (2016). Eco-ideation and eco-selection of R&D projects portfolio in complex systems industries. *Journal of Cleaner Production, 112*(5), 4329–4343.
9. VITO. (2019). *GreenSat TN4 – Assessment of selected ecodesign options*. VITO.
10. European Space Agency. (2016). *Space system Life Cycle Assessment (LCA) guidelines*. European Space Agency.
11. United Nations. (2017). *Guidelines for the long-term sustainability of outer space*. Committee on the Peaceful Uses of Outer Space.

Open Access This chapter is licensed under the terms of the Creative Commons Attribution 4.0 International License (http://creativecommons.org/licenses/by/4.0/), which permits use, sharing, adaptation, distribution and reproduction in any medium or format, as long as you give appropriate credit to the original author(s) and the source, provide a link to the Creative Commons license and indicate if changes were made.

The images or other third party material in this chapter are included in the chapter's Creative Commons license, unless indicated otherwise in a credit line to the material. If material is not included in the chapter's Creative Commons license and your intended use is not permitted by statutory regulation or exceeds the permitted use, you will need to obtain permission directly from the copyright holder.

The "Environmental Activation Energy" of Modularity and Conditions for an Environmental Payback

Karsten Schischke, Marina Proske, Rainer Pamminger, Sebastian Glaser, Nils F. Nissen, and Martin Schneider-Ramelow

Abstract Similar to the meaning of "activation energy" in physics and chemistry, there is a certain environmental investment needed for some circular design approaches: On the example of modular mobile devices, the additional environmental impact of implementing "modularity" is explained. This additional impact can be overcompensated through lifetime extension effects, if the design and related business models trigger the intended circularity effect. The paper systematically categorizes the different variants of modularity, explained on the example of smartphones. Each modularity approach features specific circularity aspects, including repair, upgrade, customization as a means to not over-spec a product, reuse and repurposing of modules. These life cycle management aspects are discussed on the example of various smart mobile products.

1 Introduction

Activation energy is the energy which must be provided to trigger a chemical reaction. Similarly, to foster a better environmental life cycle performance of a product in most cases, an additional initial manufacturing effort is needed in support of a circular design: Increased reliability might require high-quality materials, better

K. Schischke (✉) · N. F. Nissen
Fraunhofer Institute for Reliability and Microintegration (IZM), Berlin, Germany
e-mail: karsten.schischke@izm.fraunhofer.de

M. Proske · M. Schneider-Ramelow
Fraunhofer Institute for Reliability and Microintegration (IZM), Berlin, Germany

Technische Universität Berlin, Research Center Microperipheric Technologies, Berlin, Germany

R. Pamminger · S. Glaser
Technische Universität Wien, Vienna, Austria

© The Author(s) 2022
Z. S. Klos et al. (eds.), *Towards a Sustainable Future - Life Cycle Management*,
https://doi.org/10.1007/978-3-030-77127-0_2

robustness can be achieved with higher material intensity, and reparability requires a modular instead of a monolithic design. Also design for recycling might require initially design changes, which do not reduce manufacturing impacts – but are supposed to reduce impacts at end-of-life significantly. Only the use of recycled materials, as a circular design approach, tends to reduce environmental impacts right in the production phase. Figure 1 shows the comparison of an iPad with a mobile computer, which was designed with several circular design strategies in mind, such as:

- Better compatibility with accessories (see the nine-pin serial port and the RJ45 Ethernet connector, which are not found in conventional tablet computers).
- Exchangeable connector blends to allow for a shell reuse in case of changing internal electronics.
- Wood as sustainable material, which does not allow similarly small form factors as metals.
- Reparability and replaceable battery, etc.

For a more comprehensive overview of design features of this mobile computer, see Ospina et al. [1]. What is evident is the significantly larger form factor, more total material use of the circular design approach.

With this in mind, our research suggests the term "environmental activation energy" to illustrate the fact that circular design requires additional efforts – and bears the risk that these additional efforts might not pay off as expected in later product life cycle phases. Our research focusses on several examples of modular

Fig. 1 Mobile computer designed by an SME following circular design principles compared to an iPad (fifth generation)

design, as one prominent circular design approach in support of better reparability, reusability, upgradeability and recyclability.

2   Life Cycle Assessment of Modularity

With a range of examples from latest design research, the environmental impacts of circular design strategies and modularity in particular are explained to road-test our thesis that modular design comes at the cost of an "environmental activation energy".

2.1   Smartphone Modularity

The Fairphone is the most prominent example of a modular smartphone designed for do-it-yourself repairs. The Fairphone 2 launched in 2015 featured larger internal modules, mainly connected with spring-loaded connector arrays. These are robust connectors which can also withstand a rude handling by the user. These gold-coated connectors, additional printed circuit board area for contacts and module housing all resulted in a significantly higher environmental impact than a comparable conventional design [2–4].

Depending on the impact category, the impact share of modularity components is between 2.2% and 12.9% (Table 1). System boundaries are cradle to readily manufactured phone.

These additional impacts need to be compensated by the effect of a longer product lifetime due to enhanced reparability. Proske et al. [4] calculated a significantly improved environmental footprint in case this measure increases the product lifetime from in average 3 years to 5 years. This takes into account also repairs and battery replacements.

The next generation of the Fairphone launched in 2019 [5] addressed this aspect of a modularity overhead by changing the connector concept towards mezzanine strip connectors, which require only a small additional PCB footprint and feature a smaller contact area, thus less gold-coated surface finishes (Fig. 2). Some connections from the core module now had to be made with flex PCBs to bridge distances.

Table 1 LCA results: Fairphone 2 modularity (cradle to gate)

	Fairphone 2	Modularity components
Global warming (kg CO_2e)	35.16	0.77 (2.2%)
Resource depletion (abiotic, g Sb-e)	0.788	0.102 (12.9%)
Resource depletion (fossil, MJ)	139.51	8.05 (5.7%)
Human toxicity (g DCB-e)	8.290	280 (3.4%)
Ecotoxicity (g DCB-e)	110	5.79 (5.3%)

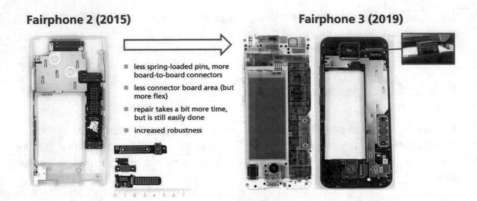

Fairphone 2 (2015)   **Fairphone 3 (2019)**

- less spring-loaded pins, more board-to-board connectors
- less connector board area (but more flex)
- repair takes a bit more time, but is still easily done
- increased robustness

Fig. 2 Evolution from modular fairphone 2 to modular Fairphone 3 and major design changes

This might affect the manufacturing impact adversely. It remains to be seen, by how much these design changes reduce the modularity-related environmental impacts and thus the "environmental activation energy", but the tendency definitely is positive.

A life cycle assessment study for the Fairphone 3 is currently work in progress. Results are expected mid-2020.

2.2 Digital Voice Recorder Concept DPM D4R

Professional digital voice recorders cover a wide range of functions. Not only the basic function "voice recording", but much more subsequent processing of the recorded files like voice recognition, creating and editing documents, adding and managing additional information and supporting the workflow via cloud solutions are in the focus. Such devices are designed for professional use in the hospital sector, by lawyers or notaries. Enabling those functions, digital voice recorders have a design similar to that of today's average smart mobile product.

A modular concept of a Digital Pocket Memo (DPM) has been designed together with a new B2B rental service [6]. This intended business model opens the possibility to replace old products with refurbished ones, update or just repair them. This leads to a lifetime extension of the whole product or single modules, and the overall life cycle impact decreases.

With this product concept, a D4R modularity approach (D4R means *design for repair*, *reuse*, *remanufacturing* and *recycling*) was applied. The product's modules are defined by components with similar end-of-life strategies (Fig. 3).

This circular design approach leads to the following six modules: The shell (mainly made out of recyclable stainless steel) consists of all parts, which are in direct contact to the user. To assure a visually nice product, these parts can only be used once, and the main end-of-life strategy is *design for recycling*. The battery is

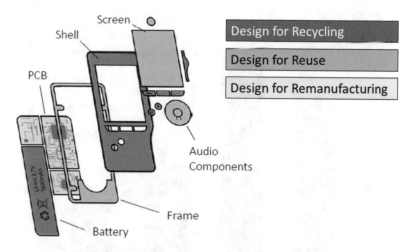

Fig. 3 Design strategies implemented with the digital voice recorder redesign

designed for recycling as the lifetime is relatively short and a certain performance is expected by a new customer. The frame (made out of recycled plastic) is the supporting structure for the PCB modules, the screen and the audio components. The frame is hidden inside (no aesthetic requirements) and meets future requirements of product updates and therefore has to be *designed for reuse*. Also, the audio module and the screen (using detachable connectors) are long lasting and are *designed for reuse*. As the environmental impact of the PCB assembly is very relevant, it should be reused. But due to short innovation cycles, the whole PCB assembly cannot be reused. Instead, the PCB itself is split into functionally grouped modules with the advantage of enabling exchange of single modules during product updates, and the PCB is *designed for remanufacturing*.

Comparing a reference product like the Philips DPM8000 and the concept DPM D4R in a scenario with a linear life cycle with no real circular approach, the impact of the DPM D4R is 12% higher. This "environmental activation energy" is caused by additional manufacturing efforts, mainly the new, modular PCB design, which enables the PCB remanufacturing. If the use time is doubled (by a second user), meaning two life cycles are taken into account (including exchange of shell module and battery, assumptions for repair of broken parts and product update, including transport, etc.), the GWP can be reduced by 21% in comparison to the reference product. If three life cycles can be realised, the GWP is reduced by 35% CO_2 eq. per product cycle [6].

Figure 4 depicts clearly the "environmental activation energy": Impacts go up with the implementation of circular design strategies and go back down only with extended lifetimes. Then, however, the positive effect can be very significant. The crucial question again is if this extended lifetime is realistic or if other limiting factors, such as component obsolescence, software obsolescence and incompatibility, might limit the possibilities for further use at the end of the first product life.

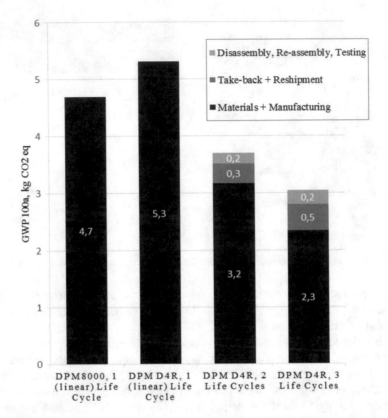

Fig. 4 Life cycle assessment results (GWP) for DVR design variants and lifetime scenarios

2.3 Embedding of Components for a Modular Printed Circuit Board Assembly

The idea of circular design has been advanced even a step further on the example of the digital voice recorder: Embedding is an advanced integration technology, where electronics components are not only placed on the surface of the PCB but are also buried in the PCB substrate. This reduces the needed area footprint for electronics modules. The PCB of the digital voice recorder is split into four distinct modules, the digital signal processor part, the internal power management, the USB connectivity and a backbone board similar to a PC mainboard [7]. The first three modules feature embedded components, and the power and USB module are assembled with non-permanent interconnection technology (screws, spring connectors, ZIF connector). This allows for a repair and refurbishment as indicated in the DVR concept outlined in the chapter before, but now with a miniaturized overall design (Fig. 5).

The image below clearly shows how complexity has been "outsourced" to the modules, featuring embedding. These modules now can be easily exchanged, easing the reuse of either modules or the backbone PCB.

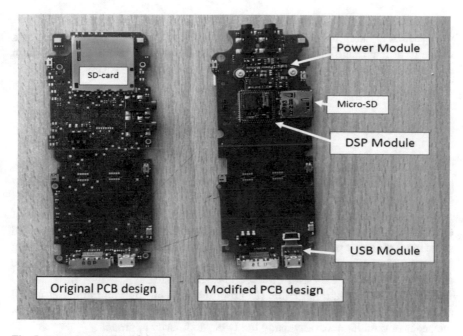

Fig. 5 Printed circuit board design changes towards modularity with embedded components

Table 2 LCA results: Digital voice recorder PCB assembly variants (cradle to gate, components excluded)

	Carbon footprint (kg CO_2 eq.)		
	Standard design	Six-layer backbone with three modules	Four-layer backbone with three modules
PCB/ backbone	1.01	1.01	0.90
USB module	–	0.17	0.17
Power module	–	0.32	0.32
DSP module	–	0.16	0.16
Connectors	–	0.007	0.007
Totals	1.01	1.67	1.57

Although Kupka et al. [8] identified a positive environmental effect of embedding as such, this does not materialize in the given design study [9]: The environmental impact is driven by the additional surface area of backbone and modules, which are – except for the DSP module – electroless-nickel/gold finishes with a high contribution to overall impacts (Table 2). Although not implemented, it seems feasible to reduce the layer count for the backbone from six to four layers. As the backbone board was not miniaturized, but defined by the existing physical dimensions of the handset, the potential of embedding is not fully exploited in this case.

As with the other examples, modularity comes at an initial environmental investment, which is likely to pay off only through lifetime extension of the device as a
whole or at least high-impact key components. In mobile electronics, these high
impacts in most cases are related to processors and memory (RAM or flash).

3 Design Rules for Modularity

The findings from the modularity assessments indicate how important it is to implement a circular design in a thought-through way and that modularity serves a well-
defined (circularity) purpose.

Usually, smart mobile devices get defective caused by a failure or damage of
only one single part, although all other parts are still working. These parts could
serve more than one product lifetime. To continue the reuse of those parts, mixing
different end-of-life strategies in one product is needed; the product's modules are
defined by its components with similar end-of-life strategies.

The following design guideline (as proposed in detail by Pamminger et al. [6]
and depicted in Fig. 6) shows roughly how to design a modular product that meets
the needs of circular economy using the D4R modularity approach.

Task 1 – Definition of the Product's CE-Strategy
The first task is to find an adequate main CE-strategy. There are four end-of-life
strategies (repair, reuse, remanufacturing, and recycling) to close the circle. The
choice depends on different aspects like how does the current business model look
like and what are the customer's needs and does my product contain valuable parts

Fig. 6 Stepwise design
approach for modular
product designs

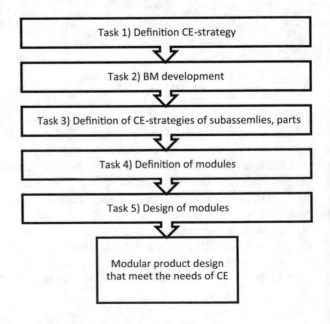

from an environmental perspective or product lifetime vs. product use-time considerations. By defining the main CE-strategy, a general direction is set for the second task, the business model development.

Task 2 – Business Model Development

A circular product design can only realise its full potential with an appropriate business model. A linear business model, which might represent the status quo, needs to be adopted to fulfil supplementary needs: reverse logistic, additional products like spare parts, new services, new activities, etc.

Tools like the CE Strategist [10] offer great help in developing circular business models.

Task 3 – Definition of CE-Strategies of Subassemblies and Parts

In the third task, the product is investigated at the component level. Depending on attributes like environmental impact, value, function, size or lifetime, the main components and parts can be identified.

For each of the main components or parts, an end-of-life strategy has to be defined, similar as with the main end-of life-strategy of the product (Task 1), but with the additional requirement, that those sub-strategies have to serve of course the product's main strategy.

Influencing factors for selecting the right strategy are the lifetime and wear, the environmental impact and the value. They are also caused by the previously defined circular business model. As with the product's main end-of-life strategy, the hierarchy of CE-cycles should receive attention.

Task 4 – Definition of Modules

This task represents the original idea of modularisation. Components, parts and subassemblies have to be clustered to modules with similar properties, end-of-life strategies, technical possibilities (interfaces, etc.) and requirements derived by the products use or the business model. A reasonable granularity should be achieved without a too detailed modularity, since a too detailed modularity will cause negative effects regarding product design, environmental impact, assembly, all sort of costs and failure susceptibility.

Task 5 – Design of Modules

The last task includes creating the module's technical structure. Questions which arise at this stage are, for example, how are the modules connected to each other, or how do the electronic interfaces look like? Can an easy and non-destructive separation of modules, which are meant for reuse or remanufacturing, be realised? Focus on design rules like "Design for Manufacturing" and "Design for Assembly" helps achieving an appropriate design. Also, automated disassembly will reduce costs with the right quantities. Include possibilities for failure detection to ensure that modules which will be reused work properly. So they can be taken again for a second life without any concerns, a convenient quality can be achieved. For a non-destructive disassembly, easy separation of modules for reuse or remanufacturing is important. In contrast, modules "designed for recycling" could be possibly removed in a destructive way (e.g. milling the housing or drilling clips or screws). When designing modules for recycling, select proper material combinations which ease the recycling process, or enable a good separability.

4 Conclusions

The comparison of modularity approaches shows the broad variety circular design strategies can have even for a rather narrow product segment: smart mobile devices.

The "environmental activation energy" is higher for those products which are built for end-user interaction, such as the DIY repair approach of the Fairphone 2 or a mix-and-match approach of functional modularity, than for those which follow, e.g. the serviceability approach only [2], where connectors do not need to withstand laymen's interaction. The potential environmental payback however is the highest, where the product remains in the hands of the end-user for a repair or even upgrade. However, also business models, which are built on modularity in a business-to-business market, can yield significant environmental savings over the lifetime. Some modularity concepts are at risk not to contribute to circularity at all, but have an adverse environmental impact over the full product life cycle: Where modularity is likely to trigger major rebound effects, the overall life cycle impact is likely to increase on top of the "environmental activation energy" of modularity. It is therefore of high importance to get clarity on the circular economy strategy and to implement appropriate design strategies stepwise, as outlined in this research.

This discussion on modularity and related environmental life cycle impacts is meant to contribute to a better understanding of the right drivers for more sustainable product concepts and factors fostering those developments.

Acknowledgement This research is part of the project sustainablySMART (www.sustainably-smart.eu) which has received funding from the European Union's Horizon 2020 research and innovation programme under grant agreement no. 680640.

References

1. Ospina, J., Maher, P., Galligan, A., Gallagher, J., O'Donovan, D., Kast, G., Schischke, K., & Balabanis, N. (2019). Lifetime extension by design and a fab lab level digital manufacturing strategy: Tablet case study. In *Proceedings of the 3rd product lifetimes and the environment PLATE conference*. Fraunhofer IZM and TU Berlin.
2. Schischke, K., Proske, M., Nissen, N. F., & Schneider-Ramelow, M. (2019). Impact of modularity as a circular design strategy on materials use for smart mobile devices. *MRS Energy & Sustainability, 6*, E16. https://doi.org/10.1557/mre.2019.17
3. Hebert, O. (2015, June 16). *The architecture of the Fairphone 2: Designing a competitive device that embodies our values*. Available at: https://www.fairphone.com/en/2015/06/16/the-architecture-of-the-fairphone-2-designing-a-competitive-device-that-embodies-our-values/. Accessed 15 Aug 2019.

4. Proske, M., Clemm, C., & Richter, N. (2016, November). *Life cycle assessment of the Fairphone 2 – Final report*. Fraunhofer IZM.
5. Christian Clemm. *The new fairphone 3 – First look with Fraunhofer IZM*. https://www.linkedin.com/pulse/new-fairphone-3-first-look-fraunhofer-izm-christian-clemm/. Accessed 14 Jan 2020.
6. Pamminger, R., Glaser, S., Wimmer, W., & Podhradsky, G. (2018). Guideline development to design modular products that meet the needs of circular economy. In *Proceedings of CARE innovation conference*. CARE Innovation Europe.
7. Manessis, D., Schischke, K., Pawlikowski, J., Krivec, T., Schulz, G., Podhradsky, G., Aschenbrenner, R., Schneider-Ramelow, M., Ostmann, A., & Lang, K.-D. (2019). Embedding technologies for the manufacturing of advanced miniaturised modules toward the realisation of compact and environmentally friendly electronic devices. In *Proceedings of EMPC 2019—22nd European microelectronics packaging conference*. Institute of Electrical and Electronic Engineers (IEEE).
8. Kupka, T., Schulz, G., Krivec, T., & Wimmer, W. (2018). Modularization of printed circuit boards through embedding technology and the influence of highly integrated modules on the product carbon footprint of electronic systems. In *Proceedings of CARE innovation 2018*. CARE Innovation Europe.
9. Schischke, K., Manessis, D., Pawlikowski, J., Kupka, T., Krivec, T., Pamminger, R., Glaser, S., Podhradsky, G., Nissen, N. F., Schneider-Ramelow, M., & Lang, K.-D. (2019). Embedding as a key board-level technology for modularization and circular design of smart mobile products: Environmental assessment. In *Proceedings of EMPC 2019—22nd European microelectronics packaging conference*. Institute of Electrical and Electronic Engineers (IEEE).
10. CE Strategist web tool, [Online]. Available: https://tools.katche.eu/strategist/. Accessed 14 Jan 2020.

Open Access This chapter is licensed under the terms of the Creative Commons Attribution 4.0 International License (http://creativecommons.org/licenses/by/4.0/), which permits use, sharing, adaptation, distribution and reproduction in any medium or format, as long as you give appropriate credit to the original author(s) and the source, provide a link to the Creative Commons license and indicate if changes were made.

The images or other third party material in this chapter are included in the chapter's Creative Commons license, unless indicated otherwise in a credit line to the material. If material is not included in the chapter's Creative Commons license and your intended use is not permitted by statutory regulation or exceeds the permitted use, you will need to obtain permission directly from the copyright holder.

Quantitative Environmental Impact Assessment for Agricultural Products Caused by Exposure of Artificial Light at Night

Yoko Kurahara and Norihiro Itsubo

Abstract Increase in artificial lighting at night adversely affects human activities, wild animals, plants, agricultural crops, and livestock. The Ministry of the Environment defines such adverse effects as "light pollution." Rice is an agricultural crop subject to the influence of light environment. We used LED lighting rice plants ("Koshihikari" cultivar) grown in a paddy field owned by professional farmers for illumination during the night and evaluated its impact on the rice's heading and yield by actual measurement. We also factored in the roadway light installed in the paddy field's vicinity and evaluated its effects on yield. Damage coefficients of light pollution for rice cultivation were developed, 18.9 $g/m^2/lx$ (equivalent to 0.046 US\$/m^2/lx) for natural white lighting and 16.4 $g/m^2/lx$ (equivalent to 0.039 US\$ /m^2/lx) for light bulb-colored lighting.

1 Introduction

The increased use and scope of night illumination reportedly resulted in an annual 2% expansion of the outdoor area illuminated by artificial lighting around the world over 16 years from 2012 [1].

According to the Ministry of the Environment, the term "light pollution" is defined as an impediment caused by light leakage from outdoor illumination or other forms of lighting and collectively the adverse effects resulting from such an impediment. Reportedly, light pollution affects human activities, wild flora and fauna, and agricultural crops in various ways. Rice is a crop subject to the influence of light pollution [2].

Rice is a short-day crop that promotes flowering beyond a certain critical dark period. Conversion of light receptor phytochrome B (phyB) between Pr-type (red

Y. Kurahara (✉) · N. Itsubo (✉)
Department of Environmental and Information Studies, TCU Tokyo City University, Yokohama, Japan
e-mail: itsubo-lab@tcu.ac.jp; itsubo-n@tcu.ac.jp

© The Author(s) 2022
Z. S. Klos et al. (eds.), *Towards a Sustainable Future - Life Cycle Management*,
https://doi.org/10.1007/978-3-030-77127-0_3

light-absorbing type) and Pfr-type (far-red light-absorbing type) is known to be involved in the expression of *heading day 3a (Hd3a)*, a flower induction gene in rice. Normally, plants in places without outdoor illumination absorb red light by photoreaction during day to induce conversion to Pr, thus increasing Pfr amount. Conversely, Pfr decreases at night, promoting *Hd3a* expression. However, it is evident that illumination by outdoor lighting and other light sources during night prevents Pfr from decreasing. Consequently, *Hd3a* expression is suppressed at the transcription level, resulting in delay in or inhibition of heading [3]. For rice of the "Hinohikari" cultivar, illumination above 2 lux (lx) caused a delay in heading; the impact was more prominent when fluorescent mercury lighting, rather than LED lighting, was used, thus suggesting differences in effects depending on the type of light source [4].

Presently, no studies have evaluated the effects of light pollution by the methodology of life cycle assessment (LCA). However, in recent years, the following items have been added as new LCA impact assessment indices: "light pollution," "ecological light pollution," and "artificial light emission." This highlights the importance of evaluating the impact of these new items on the ecosystem and human health [5, 6]. Cucurachi et al. [5] discussed the functional units of light pollution in life cycle impact assessment (LCIA) model. Given the aspects of biodiversity, they reported that not only illuminance (common measurement unit: lx) at a certain place but also light intensity and wavelength are important factors in the functional unit. Therefore, they recommended that joule (J) factoring in electrical power per unit time (watt [W]) be used as the elementary flow and argued that LCIA may be applicable to impact assessment [5].

In Japan, many papers have been published regarding the quantitative assessment results of studies targeting rice. These studies have demonstrated that an increase in the illuminance of night illumination (lx) undoubtedly adversely affects rice yield. Cucurachi et al. [5] reported the possibility of developing a model for further endpoint type damage assessment.

In this study, we evaluated the impact of night LED illumination on the heading and yield of "Koshihikari" rice grown in rice paddies owned by professional farmers. Additionally, factoring in the influence of light leakage from roadway lights installed in the vicinity of the paddy field, we evaluated the effect of light on yield and developed damage coefficients of light pollution for rice cultivation. These findings should serve as a ground for determining how to design and where to install lighting to suppress light pollution.

2   Methods

2.1   Development of a Method for Assessing Light Pollution Impact on Rice Cultivation

2.1.1   Procedures for Assessing the Impact of Light Pollution on Rice Cultivation

Referring to the impact assessment method by Itsubo et al. [7], we developed a calculation flow for light pollution damage factors based on the impact [7]. Coefficients were calculated for each of the following three categories: inventory, impact, and damage analysis. Inventory (*Inv*) was the illumination per unit of area relative to emission from one of the light source units such as outdoor lighting. Influence coefficient (*EF*) was the delay (days) in heading by increased illuminance and the impact on yield caused by the delay. Damage coefficient ($DF_{\text{light pollution}}$) was the decrease in yield and loss of profits by increased illuminance. This allows the damage on yield to be quantified when a new light source is added (Fig. 1).

$$\text{Light pollution impact} = \Sigma \, \text{inventory}\left(\text{lx} \, / \, \text{m}^2\right) \times \text{damage coefficient}\left(\text{US\$} \, / \, \text{lx or g} \, / \, \text{lx}\right)$$

(1)

2.1.2   Effect Analysis

At night, rice of the "Koshihikari" cultivar was illuminated with bulb-colored and natural white LED lights, and their effects on heading and yield were evaluated. Based on the actual measurement data, the correlations between the "illumination and delay in heading" and the "delay in heading and yield of unpolished rice" were quantified to calculate the *EF*. The cultivar used was the "Koshihikari" (*Oryza sativa* L. cv. "Koshihikari"). To explore the impact under a more realistic environment, we conducted the experiment in an approximately 20-a paddy field owned by a professional farmer in Joso, Ibaraki Prefecture, in Japan. On April 30, 2019, rice seedlings were mechanically transplanted at a cultivation density of 18 bundles per m-2 (one bundle comprising four to six seedlings). For fertilization, only basal fertilizer was applied (nitrogen, 3 kg/10 a; phosphate, 3 kg/10 a; and potassium, 3 kg/10 a). No additional fertilizer was applied, and the rice was cultivated in a conventional manner. Six light units were used (three natural white LED lights and three bulb-colored LED lights, LDR11N-W 9 and LDR11L-W 9, respectively, Ohm Electric, Tokyo, Japan). The illumination period lasted from May 4 after transplantation to September 8, the day of harvest, during which the rice was continuously illuminated during night from sunset to sunrise.

The delayed time of heading was macroscopically determined. The date when all rice plants in the plot had more than 50% of valid stalks bearing ears was regarded

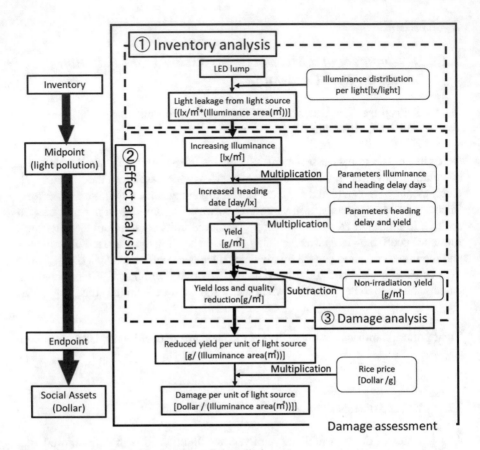

Fig. 1 Damage to rice cultivation due to light pollution

as the date of heading. The rice plants were harvested on September 8 to coincide with the harvesting period of the non-illuminated plot (Fig. 2).

2.1.3 Damage Analysis

Based on the *EF*, a relational expression for illuminance and yield decrease was calculated, which was multiplied by the relative transaction price of rice to calculate the *DF* unit [8].

Fig. 2 Lighting position
and irradiation range

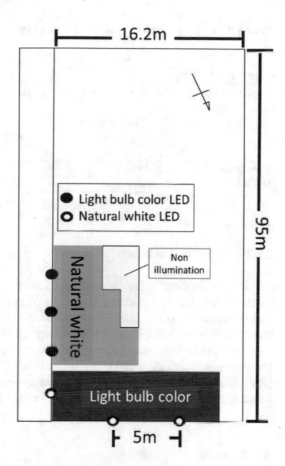

2.2   Case Study Using One Unit of Outdoor Lighting

2.2.1   Assessment Method

A case study was conducted using the damage coefficient developed in Sect. 2.1. The impact of light pollution when outdoor lighting was installed in the vicinity of the paddy field was assessed. The light used was one unit of outdoor lighting equipment (e.g., roadway/street lamps). Calculation scenario was set according to the illumination design and placement conditions satisfying the Japan Industrial Standard "JIS Z 9111 Road Lighting Standard." The situation was assumed in which a roadway/street lamp is installed near the paddy field so that crops are most susceptible to light pollution [9]. Some street lamps were inverted cone type, and others were ball-shaped, thus different in shape.

Table 1 Lighting equipment and installation conditions

Inventory data items		Unit	Street light (Inverted cone type)	Street light (ball-shaped)
Lighting equipment	Instrument luminous flux	lm	5955	6625
	Power consumption	W	60	60
	Color temperature	K	5000	5000
	Color rendering index	Ra	70	70
Installation conditions	Light height	m	4.5	4.5
	Mounting angle	°	0	0
	Maintenance rate		0.64	0.64
	Distance from object	m	0	0

2.2.2 Inventory Analysis

The *IF* was the illuminance distribution emitted per one light source unit. The illuminance distribution was calculated using three-dimensional illuminance calculation software "DIALux" [10]. The items necessary for calculating the illuminance distribution were treated as the inventory analysis parameters. Therefore, the adopted parameters were as follows: light source specifications (luminous flux [lm], power consumption [W], and color temperature [K] or color rendering properties [Ra]) and installation conditions (lamp height [m], installation angle [°], maintenance rate, and distance from the object affected by light pollution [m]) (Table 1).

3   Results

3.1   Development of Impact Assessment Methods

3.1.1   Correlation Between Illuminance and Heading Delay

Figure 3 shows the correlation between illuminance and delay in heading. The delay in heading refers to the number of days by which heading was delayed compared with the date of heading in the non-illuminated plot, which was July 26, 2019. As the illuminance increased, the date of heading was delayed. In both natural white illumination and bulb-colored illumination plots, a strong correlation was observed ($r^2 = 0.85, 0.97$). This result was consistent with that reported by Harada et al. [11]. Figure 4 shows the heading delay caused by the experiment. It is affected in a circle along the illuminance distribution. The center of the circle is irradiated with about 300 lx.

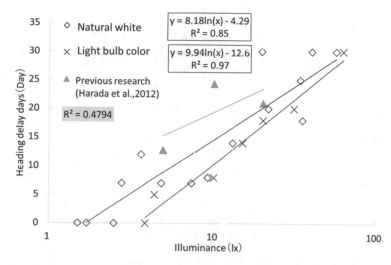

Fig. 3 Relationship between illuminance and heading delay days

Fig. 4 Paddy field affected by light on August 26

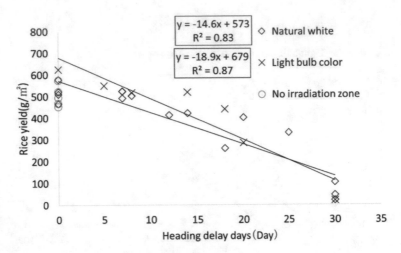

Fig. 5 Relationship between heading delay days and rice yield

3.1.2 Correlation Between Heading Delay and Yield

Figure 5 shows the correlation between heading delay and yield. The yield of polished brown rice decreased as the delay in heading increased, thus exhibiting a strong correlation ($r^2 = 0.83, 0.87$). In particular, the delay beyond 10–15 days considerably reduced the yield. Compared with the non-illuminated plot ($p = 503$), both illuminated plots (natural white and bulb-colored) exhibited a significant difference at a significance level of 1%.

3.2 Damage Factor Results

3.2.1 Correlation Between Illuminance and Yield

Figure 6 shows the correlation between illuminance and yield decrease. Consequently, the following damage coefficients were obtained: (i) natural white DF, 18.9 g/lx, and (ii) bulb-colored DF, 16.4 g/lx. Decrease in yield was observed from approximately 2 lx with natural white light and approximately 4 lx with bulb-colored light. Given the average yield in Ibaraki Prefecture ($p = 524$), the illuminance that results in zero yield was calculated to be approximately 28 lx with natural white light and approximately 32 lx with bulb-colored. Because this slope has different effects depending on the illuminance, overestimation may occur under approximately 5–15 lx, which is a limitation of this study.

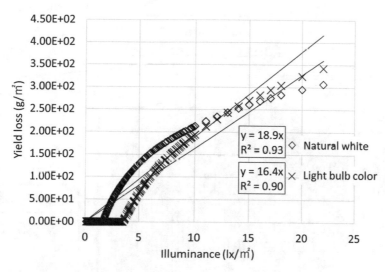

Fig. 6 Relationship between illuminance and yield

3.3 Case Study Results by One Outdoor Lighting Unit

3.3.1 Illumination Distribution in the Paddy Field

The illumination distribution in the paddy field was calculated. The values in the paddy field signified the illuminance per mesh per 1 m² (lx/m^2). The maximum illuminance of inverted cone-type street lights exceeded 20 lx and that of the ball-shaped street lights was around 6 lx. The spread of the illuminance greatly varied depending on the shape (Figs. 7, 8).

3.3.2 Endpoint Calculation Results

Figure 9 shows the decrease in the yield (damage amount) per light annually. When installed in the vicinity of the paddy field, the inverted cone-type street lights reduced the yield by 23 kg, whereas the yield reduced by the ball-shaped street lights was 10 kg. Both types of light exhibited light pollution effects. However, the effects varied depending on the distance from and the shape of the light source (due to differences in luminous flux beneath the light source). Therefore, places/situations susceptible to light pollution require countermeasures, such as installation of light shielding plates and changes in illumination position (Fig. 9).

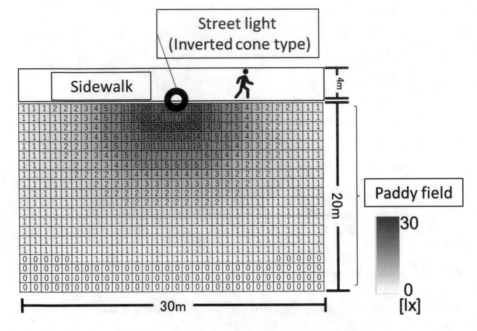

Fig. 7 Illuminance distribution result of street light (inverted cone type)

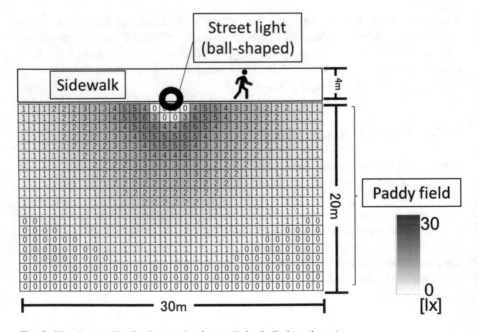

Fig. 8 Illuminance distribution result of street light (ball-shaped type)

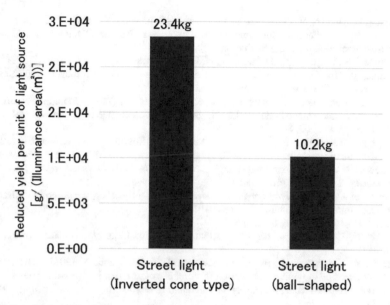

Fig. 9 Result of yield reduction per light/year

4 Conclusion

This study has proposed a framework for light pollution assessment. Additionally, we developed damage coefficients intended for assessing the impact of LED lighting on rice cultivation. Based on these damage coefficients, we conducted a case study of outdoor illumination installed in the vicinity of a paddy field. By actual measurement, a delay in heading occurred with approximately 5 lx of light. When the heading delay exceeded 10–15 days, the yield was greatly affected. Because we used data obtained from experiments, the representativeness of these results may be low. In the future, damage coefficients for assessing the ecosystem and human health should be developed.

References

1. Kyba, C. C., Kuester, T., De Miguel, A. S., Baugh, K., Jechow, A., Hölker, F., Bennie, J., Elvidge, C. D., Gaston, K. J., & Guanter, L. (2017). Artificially lit surface of Earth at night increasing in radiance and extent. *Science Advances, 3*(11), e1701528.
2. Ministry of the Environment Government of Japan, Light pollution control guidelines. http://www.env.go.jp/air/life/light_poll.html. Accessed 21 Feb 2020.
3. Ishikawa, S., Maekawa, M., Arite, T., Onishi, K., Takamure, I., & Kyozuka, J. (2005). Suppression of tiller bud activity in tillering dwarf mutants of rice. *Plant and Cell Physiology, 46*(1), 79–86.

4. Harada, Y., Kaneko, N., Haruhiko, Y., Iwaya, K., & Sonoyama, Y. (2014). The effect of LED illumination at night on heading time and yield in *Oryza sativa* L.cv. "Hinohikari,". *Journal of the Illuminating Engineering Institute of Japan, 98*(2), 74–78.
5. Cucurachi, S., Heijungs, R., Peijnenburg, W. J. G. M., Bolte, J. F. B., & De Snoo, G. R. (2014). A framework for deciding on the inclusion of emerging impacts in life cycle impact assessment. *Journal of Cleaner Production, 78*, 152–163.
6. Winter, L., Lehmann, A., Finogenova, N., & Finkbeiner, M. (2017). Including biodiversity in life cycle assessment–State of the art, gaps and research needs. *Environmental Impact Assessment Review, 67*, 88–100.
7. Itsubo, N., Murakami, K., Kuriyama, K., Yoshida, K., Tokimatsu, K., & Inaba, A. (2018). Development of weighting factors for G20 countries—Explore the difference in environmental awareness between developed and emerging countries. *The International Journal of Life Cycle Assessment, 23*, 2311–2326.
8. Ministry of Agriculture. *Forestry and Fisheries of Japan*. https://www.maff.go.jp/j/seisan/kei-kaku/soukatu/aitaikakaku.html. Accessed 21 Feb 2020.
9. JIS Z 9111- 1988 Lighting for Roads.
10. DIAL GmbH. *DIALux 4*. https://www.dial.de/en/dialux-desktop/download/. Accessed 21 Feb 2020.
11. Harada, Y., Yamamoto, H., Iwaya, K., Kaneko, N., & Sonoyama, Y. (2012). The effect of LED illumination at night on expression of floral activator Hd3a in rice with different wavelengths and luminescence. *Journal of the Illuminating Engineering Institute of Japan, 96*(11), 733–738.

Open Access This chapter is licensed under the terms of the Creative Commons Attribution 4.0 International License (http://creativecommons.org/licenses/by/4.0/), which permits use, sharing, adaptation, distribution and reproduction in any medium or format, as long as you give appropriate credit to the original author(s) and the source, provide a link to the Creative Commons license and indicate if changes were made.

The images or other third party material in this chapter are included in the chapter's Creative Commons license, unless indicated otherwise in a credit line to the material. If material is not included in the chapter's Creative Commons license and your intended use is not permitted by statutory regulation or exceeds the permitted use, you will need to obtain permission directly from the copyright holder.

City Air Management: LCA-Based Decision Support Model to Improve Air Quality

Jens-Christian Holst, Katrin Müller, Florian Ansgar Jaeger, and Klaus Heidinger

Abstract Siemens has developed an emission model of cities to understand the root cause and interactions to reduce air emissions. The City Air Management (CyAM) consists of monitoring, forecasting and simulation of measures. CyAM model aims to provide formation on air pollution reduction potential of short-term measures to take the right actions to minimize and avoid pollution peaks before they are likely to happen. The methodology uses a parameterized life cycle assessment model for transport emissions and calculates the local impact on air quality KPIs of individual transport measures at the specific hotspot. The system is able to forecast air quality and by how it is expected to exceed health or regulatory thresholds over the coming 5 days.

In this paper, the LCA model and results from selected cities will be presented: Case studies show how a specific combination of technologies/measures will reduce the transport demand, enhance traffic flow or improve the efficiency of the vehicle fleet in the vicinity of the emission hotspot/monitoring station.

1 Introduction

According to the World Health Organization (WHO), almost 90% of the world's urban population breathe air with pollutant levels that far exceed the recommended thresholds. Approximately seven million people die each year from the effects of air pollution, which, according to the WHO, makes it a greater global health threat than Ebola and HIV [1].

City leaders are under pressure to meet these challenges and define strategies for sustainable, clean and smart growth. However, they often lack sufficient data or

J.-C. Holst (✉) · K. Müller · F. A. Jaeger
Siemens AG, Corporate Technology, Berlin, Germany
e-mail: jens-christian.holst@siemens.com

K. Heidinger
Siemens AG Austria, Siemensstrasse, Vienna, Austria

© The Author(s) 2022
Z. S. Klos et al. (eds.), *Towards a Sustainable Future - Life Cycle Management*,
https://doi.org/10.1007/978-3-030-77127-0_4

digital tools necessary to make the best decisions. Additionally, continuous urbanization has resulted in population growth, sprawling land use and changes in mobility behaviour. Despite public transit investments, congestion is worsening globally. The sheer volume of inter- and intra-urban transportation has outpaced improvements in and customer uptake of clean transport technology. As a result, air quality has deteriorated in many cities, large and small, and city leaders are accepting that, at its core, poor air quality is an issue of public health and wellbeing [2].

As trusted global partner for sustainable city development, Siemens has developed a complete, cloud-based software suite to overcome the challenges of poor air quality using artificial neural networks and LCA-based decision support methodology. The City Air Management Tool visualizes air quality data recovered from municipal measuring stations in real time. In addition, it forecasts air pollution levels for the next 3–5 days with up to 90% accuracy and also simulates the impact of short-term measures on air quality. Combining air quality forecasts with the simulation of the effectiveness of planned measures and technologies helps cities in the first instance to activate short-term measures; however, it will also foster long-term air quality improvement measures in the upcoming years, such as the implementation of low emission zones or increased e-mobility.

2 City Air Management: Solution and Methodology

The process of CyAM is depicted in Fig. 1. The system starts with data collection from measuring station at hotspots in cities [3]. Then a simulation predicts the degree of air pollution several days in advance. Based on its analytical capacities, the main drivers for air pollution are identified and monitored continuously in order to improve the prediction capabilities. The software aims to give cities the information needed to minimize and avoid pollution peaks before they are likely to happen.

The main technical specifications of the CyAM are as follows:

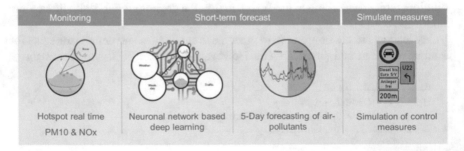

Fig. 1 City air management process: monitoring of air quality KPIs, forecast and simulation of applicable measures

- Monitor the citywide, hotspot emissions of all environmental sensors which have been integrated in the tool, focusing primarily on PM2.5, PM10 and NOx. Data is shown for each sensor on an hourly basis.
- Forecast air quality and inform city leaders through a dashboard about where and by how much air quality is expected to exceed health or regulatory thresholds over the coming 3 days with 90 per cent accuracy and up to 5 days at a level of 75–80%.

Cities/counties/states operate their own air pollution sensor networks in order to prove their compliance with national or international regulations [4, 5] . This data is gathered on central severs and publicly available in most parts of the world. The city would need to provide air quality sensor data, historic and real-time air quality data streams of all available measurement stations from a central database using standard database data interfaces. The CyAM has an API, which allows this data to be pulled from these servers or pushed to the CyAM as soon as the data is available. This is depicted in Figs. 2 and 3.

The dashboard shows the three air quality KPIs, NO2, PM10 and PM2.5, over a timeline. Potential transportation-related measures to improve the air quality are shown on the right-hand side. The data for the individual measurement stations are visualized, categorized and benchmarked against the legal thresholds in the dashboard. It provides an immediate evaluation of the current situation and information on whether it is necessary to act. The latest history is also available for review, as well as the gliding annual average (Fig. 3).

There are two options to do forecasting for air pollutants, domain models and artificial intelligence [6].

Domain models are models which fully understand the physical and chemical processes of emission source behaviour and the atmospheric processes during transmission of pollutants. Their main disadvantage is that there are a vast variety of emission sources in and around a city. It is very expensive and time consuming to

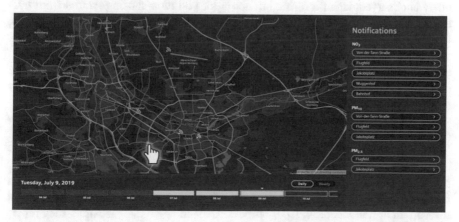

Fig. 2 Dashboard of measuring stations and potential measures to reduce emissions for KPIs like NOx and PMx

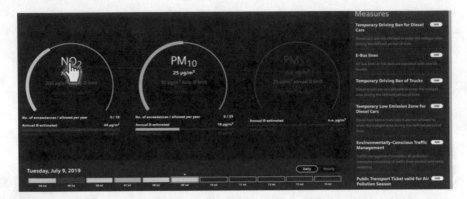

Fig. 3 Dashboard of actual air pollution data for NO2 and PM10

assess them in real time. The modelling of the transmission (distribution plus the physical and chemical processes of the pollutants in the air) is time consuming and requires high computing capacities.

CyAM uses artificial intelligence (AI) to forecast air pollution concentrations at individual air quality measurement stations [7]. It takes few available parameters which are available as forecasts and builds an empiric model based on historic data. CyAM uses air pollution measurement data, weather data/weather forecast data, calendric data and special events. The AI finds correlations and patterns in this data to predict air pollution for individual measurement stations. It doesn't contain any knowledge about the physical and chemical processes, responsible for these concentrations. Based on real-time data and forecasts of weather – and calendric/event data – a 5-day forecast is provided. The Advantage is a model which has a high precision, takes little computing power during operation and requires few data points.

In the air pollution forecasting system, recurrent neural networks are used, which are well suited for this task. They also make it easier to uncover a great deal of previously unobserved, latent information about air pollution-causing factors from traffic, industries, agriculture, etc., in the internal dynamical model of the environment, which is built up during the training of the network. Based on all of the resulting data, as well as seasonal and immediate weather forecasts, the neural network has to learn how to predict the degree of air pollution. During the city-specific training process of the system, which includes hundreds of iterations, the program steadily reduces the difference between its forecasts and the actual levels of pollutants measured in the city's atmosphere by changing the weightings of individual parameters (Fig. 4).

In order to calculate impacts of individual measures, a domain model is inevitable [8]. The key to success is to only model the share of emission and concentrations, which can actually be impacted by interventions. This reduces the data requirements and complexity to a minimum, using forecasting values for different measurement stations, provided by the AI. Depending on their location, they represent certain emission sources, as indicated in Fig. 5. If the measurement station at

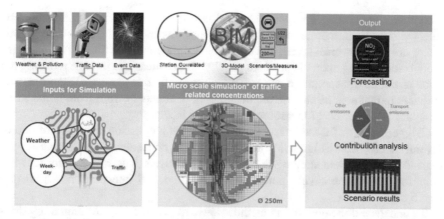

Fig. 4 Forecasting of air pollution is realized by domain models or artificial intelligence; CyAM uses the second

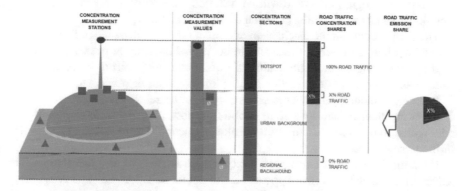

Fig. 5 Urban emission concentration profile for PM or NOx

the roadside (referred to as hotspot) and the one representing the local background are located well, they can be used to estimate the local road traffic-based concentration in a street canyon in front of the roadside measurement station. It is a simplified concentration contribution analysis, singling out local traffic-related air pollution concentration.

In parallel, a domain model is used to perform an emission contribution analysis of different vehicles and vehicle categories from the overall local transport emissions. These are based on information such as how many vehicles of which vehicle category and exhaust gas emission class are passing by the measurement station. Temperature, slope and congestion level are represented as well. The combination of the contribution model from roadside emissions and the concentration contribution analysis based on correlating forecasted air quality sensor data provides a full view of which vehicles are responsible for a certain air pollution concentration [9]. This process is executed for any hour of the forecasting period individually. The domain model for traffic emissions also contains a variety of location-specific scenarios, representing different intervention options. These interventions can be selected short term in order to reduce local emissions and thus local concentrations. There are 17 short-term levers that could be simulated within the CyAM standard model – depending upon the needs of the city. These measures include reducing the price of public transportation and encouraging public transport use, requiring that all buses in that area be electric or encouraging residents to work from home when possible. The selected 17 measures are depicted in Fig. 6.

Triggering any of these interventions results in an emission reduction of one or several modes, what is modelled and translated in concentration reductions at the roadside measurement station via contribution analysis. The result is an hour-by-hour forecast of the saving potential for the modelled interventions within the next 5 days.

To understand the underlying methodology, we use the lever of temporary driving ban for diesel cars. In case of exceeding emission limits, a diesel driving ban is announced and enforced. All diesel cars are restricted to enter the city region. A licence plate recognition system will be installed around city boundaries and near emission hotspots to check the potential driving permission [10].

The LCA model and the mechanism of demand shift are depicted in Fig. 7. After a diesel car ban is in effect, the available additional capacity at peak time will be used by increasing capacity utilization of public transport, shift to bicycle, to carpooling and even absolute reduction of car pkm due to home office, etc.

In the LCA model, passenger kilometre (pkm) of diesel cars is shifted using four different mechanism step-by-step to public transport, to bicycles, to carpooling or ride-hailing and if required to absolute pkm reduction due to home office, work shifting/vacation, etc.

From the 17 levers of our model, the two main impacts of air quality emission reduction are modal shift (from private cars to public transport, taxi and zero emission vehicles) and improved traffic flows by capacity shifting [9].

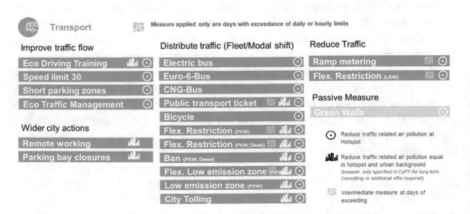

Fig. 6 Seventeen transportation measures to reduce air pollution at hotspot

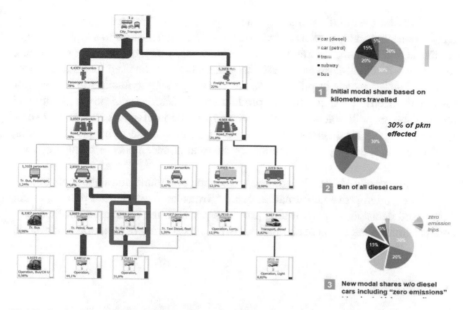

Fig. 7 Example for LCA model of urban transportation: diesel car ban reduces the emissions by modal shift and improved traffic flow

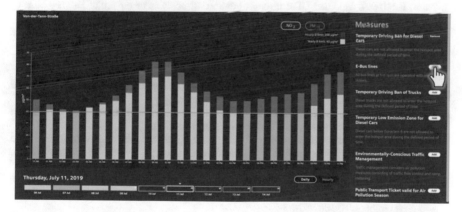

Fig. 8 Quantification of reduced NO2 value at hotspot by applying diesel ban as a measure; temporal behaviour from forecast is also shown

3 Results and Discussion

From Fig. 8, we can understand the decision support function of the CyAM system. The forecast of the emission value for hours and days will be combined with the impact evaluation of the system. By applying different levers, one can see the reduction of certain KPIs for different time scales. The shown example displays a diesel car traffic ban over certain time and the respective NO2 reduction at the hotspot. CyAM can support cities to conduct knowledge-driven decisions to avoid exceeding set limits and combine different measures to increase impact analysis and outlook for set mid- and long-term measures. The system can also be used for other applications, i.e. adjust means of private transportation by using dynamic traffic zones or restrict to electric cars, control the production of factories and power plants and monitor the air pollution nearby, plan time and location of sports and social events to ensure a good air quality and monitor air pollution close to hospitals, schools, kindergartens and living communities for early announcements or change of time and location.

In summary, CyAM can be applied to reshape communication about air quality in your city by improved information quality, transparency, measure polling and/or data-based decision support. It also provides possibilities for short-term measures and real-time management of air quality, i.e. by peak shaving or temporal modal shift.

References

1. World Health Organization. (2008). *The global burden of disease: 2004 update*. WHO. http://www.who.int/healthinfo/global_burden_disease/2004_report_update/en/
2. Zaim, K. (1999). Modified GDP through health cost analysis of air pollution: The case of Turkey. *Environmental Management, 23*(2), 271–277.
3. http://www.esa.int/esaEO/SEM340NKPZD_index_0.html; North American Space Agency http://www.nasa.gov/topics/earth/features/health-sapping.html;
4. Environmental Protection Agency. *Air quality standards*. Available at: http://www.epa.ie/air/quality/standards/. Accessed 4 Nov 2016.
5. World Health Organisation. *Air quality guidelines – Global update 2005*. Available at: http://www.who.int/phe/health_topics/outdoorair/outdoorair_aqg/en
6. Bai, L., Wang, J., Ma, X., & Lu, H. (2018). Air pollution forecasts: An overview. *International Journal of Environmental Research and Public Health, 15*, 780.
7. Schneegass, D., Udluft, S., & Martinez, T. Improving optimality of neural rewards regression for data-efficient batch near-optimal policy identification. In *ICANN 2007: 17th international conference, 2007, proceedings, part I* (pp. 109–118).
8. Jaeger, F., et al. (2017). LCA in strategic decision making for long term urban transportation system transformation. In *8th international conference LCM* (pp. 193–204). https://link.springer.com/book/10.1007/978-3-319-66981-6
9. Zhang, W., Lin Lawell, C.-Y., & Umanskaya, V. (2016, December). *The effects of license plate-based driving restrictions on air quality, theory and empirical evidence*, Clinawell. Retrieved from http://clinlawell.dyson.cornell.edu/driving_ban_paper.pdf at 31.01.2018.
10. https://assets.new.siemens.com/siemens/assets/public.1531141935.fe2fede0e226049cf69f113 7d00a386be58e79df.8063-02-onepager-city-air-mngmt-en-180709-3.pdf

Open Access This chapter is licensed under the terms of the Creative Commons Attribution 4.0 International License (http://creativecommons.org/licenses/by/4.0/), which permits use, sharing, adaptation, distribution and reproduction in any medium or format, as long as you give appropriate credit to the original author(s) and the source, provide a link to the Creative Commons license and indicate if changes were made.

The images or other third party material in this chapter are included in the chapter's Creative Commons license, unless indicated otherwise in a credit line to the material. If material is not included in the chapter's Creative Commons license and your intended use is not permitted by statutory regulation or exceeds the permitted use, you will need to obtain permission directly from the copyright holder.

Is Environmental Efficiency Compatible with Economic Competitiveness in Dairy Farming? A Case Study of 80 Luxembourgish Farms

Rocco Lioy, Caroline Battheu-Noirfalise, Aline Lehnen, Roman Reding, and Tom Dusseldorf

Abstract The aim of the study was to investigate both environmental and economic performances of Luxembourgish dairy farms in order to assess possibilities and limits of improving economic competitiveness via increasing environmental efficiency. In the environmental field, four LCA impact categories (carbon footprint, energy consumption, acidification, eutrophication) were analysed, while in the economic field, costs, incomes and profit of the farms were investigated. A main result was that a sustainable dairy production with less environmental impact in all considered categories is also of advantage in terms of farm competitiveness. The most efficient farms reach also the highest profit. The case study proves that a high environmental performance is not only of advantage in terms of economic competitiveness, but is even a necessary prerequisite for best economic performances.

1 Introduction

The case study was carried out in the frame of the Interreg VA Program of the European Union (Project AUTOPROT). This project aims to investigate if and to which extent an increase of protein self-sufficiency (autarky) can lead to a better competitiveness of dairy farms and to a reduction of their environmental impact as well. After the abolition of the milk quota system in the European Union at the end of March 2015, dairy farms were forced more than ever to increase production efficiency as a precondition to improve their own competitiveness. Thus, in the frame of this study, a combined environmental and economic analysis of dairy farms was

R. Lioy (✉) · A. Lehnen · R. Reding · T. Dusseldorf
CONVIS Société Coopérative, Ettelbruck, Luxembourg
e-mail: rocco.lioy@convis.lu

C. Battheu-Noirfalise
Centre Wallon de Recherches Agronomiques (CRA-W), Libramont, Belgium

© The Author(s) 2022
Z. S. Klos et al. (eds.), *Towards a Sustainable Future - Life Cycle Management*,
https://doi.org/10.1007/978-3-030-77127-0_5

49

carried out in order to highlight possibilities and limitations of a conciliation of environment and competitiveness in dairy farming.

2   Material and Methods

2.1   The Investigated Farms and the Protein Autarky

The investigation refers to a sample of 80 Luxembourgish dairy farms supervised in the years 2014, 2015 and 2016. The figures of crop production and animal husbandry of the investigated farms (Tables 1 and 2) as well as all the figures presented in this study refer to the average of the three investigation years. The farms (ca. 11% of all dairy farms of the land) cover the different dairy production systems in the country and are representative of dairy production in Luxembourg.

A very important indicator for the farms is the self-sufficiency degree of protein in dairy farms, in subsequently called protein autarky. There are two possibilities to express protein autarky. The first one refers to the performance of farm crop production to deliver protein for the herd. In case of crop production, the autarky is the amount of on-farm produced protein in relation to the total protein fed [1]. The other indicator of protein autarky refers to the performance of animal production to valorise protein fed. This figure takes into account the protein need based on need Tables [2] and considers as valorised the difference between needed and purchased (with concentrate and roughage) protein. The purchased protein is estimated based on feed protein Tables [3]. A detailed description of this figure is shown in [4].

2.2   The LCA Methodology Applied and Economic Indicators Used

The investigation of environmental impact was carried out on four LCA midterm impact categories (carbon footprint, energy consumption, acidification and eutrophication). The carbon footprint takes into account not only emissions deriving from production means, animal husbandry and crop production but also carbon credits deriving from humus storage into arable soils and via renewable energies

Table 1 Main figures of crop production

Figure	Average	St. dev.
Farm size (dairy production)	87.08 ha	45%
Cereals	8.25 ha	78%
Maize silage	16.39 ha	54%
Grassland (permanent + temporary)	61.89 ha	46%
Other feed plants	0.55 ha	328%

Table 2 Main figures of animal husbandry

Figure	Average	St. dev.
Animal density	1.56 LAU/ha	19%
Dairy cows	84 (n)	55%
Production intensity	7.550 kg ECM/ha	29%
Dairy performance	7.847 kg ECM/year	15%
Concentrate use	6.33 kg/cow/day	26%
Concentrate efficiency	0.29 kg/kg ECM	21%

LAU Large animal unit, ECM Energy-corrected milk

Table 3 Sources of emission and credit factors for carbon footprint

Emission or credit post	Source
Production means (manufacturing and transport)	Ecoinvent 2009 [5]
Enteric fermentation and manure management	IPCC 2006 [6]
Indirect soil emissions	IPCC 2006 [6]
Mineral nitrogen fertilisation	IPCC 2006 [6]
Fuel (manufacturing and combustion)	Ecoinvent 2009 [5]
Humus balance of arable land	Leithold et al. 1997 [7]
Electricity from biogas	Ecoinvent 2009 [5]

(Table 3). This means that the carbon footprint results in a net balance of CO_2-equivalents.

In the case of humus balance of arable soils [7], the balance results in an emission if negative, and in a credit, if positive. The global warming factors used for carbon footprint were 25 for methane and 298 for dinitrogen oxide, according to [6]. The allocation between milk and meat was carried out following their protein content.

The energy consumption (no renewable energy) was estimated by taking into account not only direct energy (fuel and electricity) but also the indirect energy for manufacturing and transport of used production means and investments (buildings and machinery). The source of these energy consumptions was the Ecoinvent data-basis [5].

Acidification takes into account the SO_2-equivalents deriving from SO_2, NH_3 and NOx. The sources for the emission factors for the three gases were in the case of used production means [5] and in the case of livestock and crop production [8] for NH_3 and [6] for NOx (as NO). The characterisation factors for NH_3 and NOx (as NO) were derived from [9].

Finally, in the case of eutrophication, the estimation of nitrate leaching was made as difference between the nitrogen balance at farm gate and the sum of all emission of N-species as well as the N-storage into the soil, in analogy to [10]. The PO_4-equivalents coming from phosphorous emission are estimated based on farm gate balance for phosphorus. Even in the case of eutrophication, characterisation factors for PO_4-equivalents from different eutrophication sources were derived from [9].

As shown by [11] and [12], the behaviour of carbon footprint when expressed in function of product (kg ECM) or farm size (ha) is contradictory. Thus, to avoid misunderstandings in interpretation of results, for all investigated impact categories, both functional units (per kg ECM and per ha) were used.

In this study, the incomes without subsidies, the production costs for a kg ECM as well as the profit (as a difference between the first two) were used as economic indicators.

2.3   The Principle of Farm Segregation and Statistics Analysis

In order to analyse result variability, according to [13], the investigated farms have been divided into groups by crossing the X and the Y axis in the average value of carbon footprint per ha (11.2 t CO2eq) and per kg ECM (1.32 kg CO2eq) (see Fig. 1).

This allows the segregation of farms into four groups, which are well differentiated in terms of production intensity and efficiency of production mean use (as will be clear more below, Fig. 4). In particular, the farms with only one indicator of carbon footprint better than the average are farms with the highest or lowest production intensity. The farms of the other two groups (with both values of carbon footprint better or worse than the average) are farms with a middle-intensive production intensity, when compared with the other two groups. Figure 1 also shows the used denomination of the four farm groups.

Concerning the statistic methodology, the analysis was carried out by using the program "R", which is freely available on the Internet [14]. ANOVA test was used for determining the significance of selected figures in the whole pool, while Tukey post hoc test was used among the segregated farm groups. The conditions of

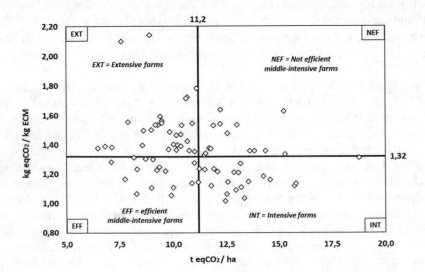

Fig. 1 The segregation and the denomination of the four farm groups

application of the ANOVA test, homogeneity and homoscedasticity, were tested using the Shapiro test and the Bartlett test, respectively. An exhaustive description of ANOVA test can be found under [15].

3   Results and Discussion

The value of protein autarky of crop production of investigated farms (Table 4) shows that two thirds of the protein fed were produced on farm, and the other third was purchased. In the case of animal production, on average the farms show a valorisation of the on-farm produced protein of 49%.

This means that roughly one half of the on-farm produced protein is lost. These losses are problematic, because they result in higher emissions (especially NH3, [16]) and in a higher import of feed (with consequent higher energy consumption and carbon footprint). In terms of variability, the purchased protein shows the maximum value. As we will see below (Fig. 5), the protein purchase plays a key role in explaining differences among the farms.

The results of LCA impact categories show a very high variability both in product (Table 5)- and in surface (Table 6)-related figures. The largest spread between minima and maxima values can be found in the eutrophication figures. The calculated figures for dairy farming in Luxembourg are consistent with the range of values from literature [17–19] concerning all product-related figures as well as surface-related figures of carbon footprint and energy consumption. Only in the case of surface-related figures of acidification and eutrophication, it was not possible to find values in the literature because relating these figures to the farm area is unusual.

In the case of economic results (Table 7), there is an evident difference between incomes and costs on the one hand and profit on the other hand. Indeed, the variability of the first two parameters is clearly lower than those for the profit, which varies very largely among the farms. In any case, on an average, the farms are capable of reaching only a very low profit, if subsidies are not considered.

Table 4 Figures of protein autarky of investigated farms

Protein autarky	Value	St. dev.
On-farm produced protein (1)	966 kg CP/ha	54%
Purchased protein (2)	497 kg CP/ha	81%
Total protein fed (3) = (1) + (2)	1.462 kg CP/ha	60%
On-farm protein autarky = (1) / (3) * 100	66%	14%
Needed protein by dairy herd (4)	982 kg CP/ha	62%
Valorised protein (5) = (4) − (2)	485 kg CP/ha	58%
CP-autarky (anim. prod.) = (5) / (4) * 100	49%	29%

CP Crude protein

Table 5 Product-related impact of farms in the investigated LCA categories

Impact category	Functional unit: 1 kg ECM	St. dev.	Min.	Max.
Carbon footprint	1.32 kg CO2eq	16%	1.02	2.14
Energy consumption	4.8 MJ	19%	3.3	8.0
Acidification	17.3 g SO2eq	21%	12.0	36.3
Eutrophication	11.7 g PO4eq	36%	6.1	29.4

Table 6 Surface-related impact of farms in the investigated LCA categories

Impact category	Functional unit: 1 ha	St. dev.	Min.	Max.
Carbon footprint	11.2 t CO2eq	21%	6.5	18.8
Energy consumption	41 GJ	27%	19	65
Acidification	148 kg SO2eq	23%	80	230
Eutrophication	99 kg PO4eq	33%	35	196

Table 7 Economic figures of investigated farms (*incomes are without subsidies*)

Economic figures	€-cent/kg ECM	St. dev.	Min.	Max.
Incomes	39.7	9%	34.3	55.7
Costs	38.8	20%	23.1	63.2
Profit (incomes-costs)	0.9	822%	−24.6	19.9

Table 8 LCA figures of segregated farm groups and range of results

LCA figure	EFF	Range	EFF	Range	EFF	Range	EFF	Range
Kg CO$_2$eq/kg ECM	1.2	2	1.17	1	1.51	4	1.45	3
t CO$_2$eq/ha	9.2	1	13.9	4	9.5	2	12.6	3
MJ/kg ECM	4.4	2	4.3	1	5.3	4	5.1	3
GJ/ha	34	2	51	4	34	1	45	3
g SO$_2$eq/kg ECM	16.6	2	15.4	1	19.5	4	18.9	3
Kg SO$_2$eq/ha	127	2	182	4	123	1	164	3
g PO$_4$eq/kg ECM	10.5	2	9.7	1	14.1	4	13.3	3
Kg PO$_4$eq/ha	81	1	115	3	88	2	116	4
Sum of ranges	–	14	–	19	–	22	–	25

The farm segregation allows ranging the results among farm groups. As can be seen in Table 8, the middle-intensive farms with a high efficiency (EFF) show the lowest environmental impact, if all the ranges in the eight impact categories are added up. In the hierarchy of the range, the group EFF is followed by the intensive farms (INT), then by the extensive (EXT) and finally by the middle-intensive farms with a low efficiency (NEF). In the case of product-related emissions, the farm group INT each times reaches the best performances, but this situation inverts when the results are related to the ha of the farm. In that case, the intensive farm group shows the weakest results in the range with only one exception (kg PO4eq/ha).

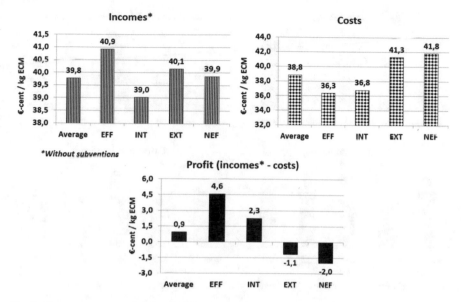

Fig. 2 Economic figures of segregated farm groups

The situation of the extensive farm group (EXT) is inverse to the intensive farms. This suggests that the farm structure is important in order to influence the range of result.

This hierarchy in environmental results among the farm groups is found to be the same also as in the case of economic results. As shown in Fig. 2, the farm group EFF reaches the best profit per kg ECM, followed by the groups INT, EXT and NEF.

It should also be noted that the farms of the group EFF have the highest value in terms of incomes and the lowest value in terms of costs, which explains the higher profit in comparison with the other groups. It is also interesting to observe that intensive farms are able to keep the costs low, but in terms of income, they reach the lowest rates. The other two groups (EXT and NEF) are not able to reach a positive profit, if subsidies are not taken into consideration.

In order to explain this hierarchy in the results, it is helpful to show the figures linked to the structure (Fig. 3) as well as to the management of the farm groups (Fig. 4). The figures of animal density as well as production intensity confirm that the groups INT and EXT have respectively the highest and the lowest production intensity and that the other two groups (EFF and NEF) are located in between, with NEF showing on average a higher intensity than EFF. The EFF group shows the lowest value in the farm area and the second lowest value in terms of number of dairy cows, very close to the lowest value of EXT group. This is a first hint that the farms of the EFF group try to capitalise maximally their own resources because these are limited in comparison with other groups. It appears consistent with the figures of animal density and production intensity that intensive farms (INT) show the highest values in farm size as well as in number of cows.

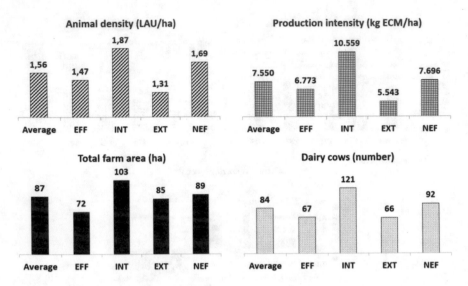

Fig. 3 Main figures of farm groups related to the farm structure

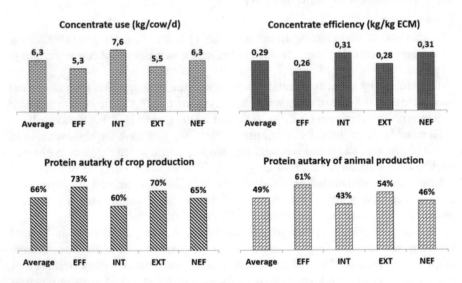

Fig. 4 Main figures of farm groups related to the feeding management

A second important point concerns the influence of management quality on the environmental impact of farm groups. As can be observed in Fig. 4, the most efficient farms (EFF) show also the best values not only in concentrate management but also in protein autarky. This is consistent with observation of other authors [20, 21], who stressed that feed management has a huge impact on the environmental result in livestock/dairy production in general and on carbon footprint in particular.

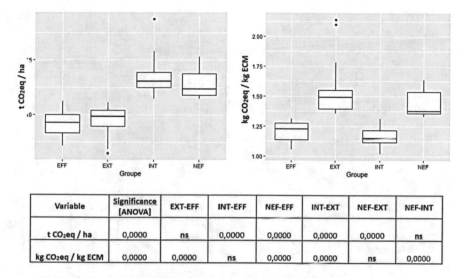

Fig. 5 Statistics of carbon footprint for the farm groups (*ns* not significant)

Variable	Significance [ANOVA]	EXT-EFF	INT-EFF	NEF-EFF	INT-EXT	NEF-EXT	NEF-INT
t CO₂eq / ha	0,0000	ns	0,0000	0,0000	0,0000	0,0000	ns
kg CO₂eq / kg ECM	0,0000	0,0000	ns	0,0000	0,0000	ns	0,0000

Further, it is interesting to point out that the hierarchy in feed management, as shown in Fig. 4, is more similar to the figures of surface-related impacts (Table 8) than to those of product-related ones. Indeed, if results are expressed per ha, extensive farms score better than intensive ones, and the latter show the worse result, which also is the case in Fig. 4. The highest level of milk production of intensive farms is evidently capable of concealing deficits in important management sectors such as feeding and providing better figures for these farms, if results are expressed related to the kg ECM.

A last consideration concerns statistical significance of differences in results among farm groups. As shown in Fig. 5 (for reasons of space, the analysis refers only to carbon footprint, but with few differences to the other three impact categories), the major part of differences among the groups are significant, although only in two situations (NEF-EFF and INT-EXT), the significance is given for both product- and surface-related indicators, which could be expected because of the kind of segregation. Further, the most efficient farms (EFF) show a behaviour that is closer to the extensive one, if the result is related to the surface, and to the intensives, if the result is related to the product. This suggests that these farms are best capable of combining a higher level of efficiency with a low level of environmental impact.

In the case of economic figures, the significance is only given for the groups EXT-EFF and NEF-EFF (Fig. 6). In the case of the pair EFF-INT, there is no significance in economic results, despite the fact that in the average the profit of the farm group EFF is higher than the group INT (Fig. 3). Nevertheless, the fact that in comparison with the other two groups (EXT and NEF) the results of the EFF group are better underlines that a better management also results in better economic figures.

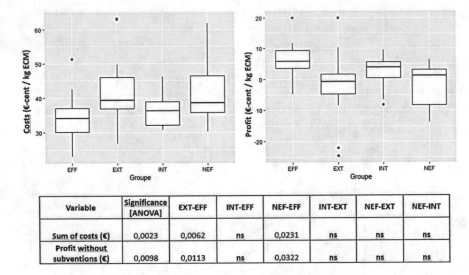

Variable	Significance [ANOVA]	EXT-EFF	INT-EFF	NEF-EFF	INT-EXT	NEF-EXT	NEF-INT
Sum of costs (€)	0,0023	0,0062	ns	0,0231	ns	ns	ns
Profit without subventions (€)	0,0098	0,0113	ns	0,0322	ns	ns	ns

Fig. 6 Statistics of economic parameters for the farm groups (*ns* not significant)

4 Main Conclusions

- The use of product- and surface-related functional units allows a better understanding of differences in results among the farms.
- In particular, differences among intensive and extensive farms are mostly due to the farm structure, those among middle-intensive ones mostly to the efficiency in farm management.
- Although not always supported by statistics, there is evidence that efficient middle-intensive farms show better performance both environmentally and economically. Their environmental efficiency allows best economic performance.
- Based on this study, the conclusion that intensive farms always show better performance because of better product-related results cannot be confirmed.
- In times of liberalisation of milk quota, a smart feed management (especially of feed protein) seems to be a key lever for realising best performances both environmentally and economically.

References

1. IDÈLE. (2010). *Guide pratique de l'alimentation du troupeau bovin laitier*. Idèle-Editions Quae.
2. GfE. (2001). *Empfehlung zur Energie- und Nährstoffversorgung der Milchkühe und Aufzuchtrinder*. DLG-Verlag.
3. DLG-Futterwerttabellen Wiederkäuer. (1997). *7. erweiterte und überarbeitete Auflage*. DLG-Verlag.

4. Lioy, R., Klöcker, D., Laugs, P., & PetrY, J. (2019). *Eiweißautarkie Luxemburger Milchviehbetriebe – Stand, Verbesserungs-potential, räumliche Variabilität.* Internationale Grünlandtagung. http://www.grengland.lu/sites/default/files/files/brochure_iglt_2019_d_f_v2_low.pdf. Accessed 15 Jan 2020
5. ECOINVENT. *The live cycle inventory data,* Version July 2009 https://www.ecoinvent.org/. Accessed 15 Jan 2020.
6. IPCC. (2006). *Greenhouse gas inventory.* Reference manual (vol. 4). https://www.ipcc-nggip.iges.or.jp/public/2006gl/vol4.html. Accessed 15 Jan 2020.
7. Leithold, G., Hülsbergen, K.-J., Michel, D., & Schönmeier, H. (1997). Humusbilanzierung – Methoden und Anwendung als Agrar-Umweltindikator. In Diepenbrock W., Kaltschmitt M., Nieberg H., Reinhardt G. (Hrsg.). *Umweltverträgliche Pflanzenproduktion – Indikatoren, Bilanzierungsansätze und ihre Einbindung in Ökobilanzen.* Zeller Verlag Osnabrück
8. Haenel, H.-D., Rösemann, C., Dämmgen, U., Poddey, E., Freibauer, A., Wulf, S., Eurich-Menden, B., Döhler, H., Schreiner, C., Bauer, B., & Osterburg, B. (2014). *Calculations of gaseous and particulate emissions from German agriculture 1990 – 2012: Report on methods and data (RMD) submission 2014.* Johann Heinrich von Thünen-Institut, 348 p, Thünen Rep 17, 2014 https://www.thuenen.de/media/publikationen/thuenen-report/Thuenen-Report_17.pdf. Accessed 15 Jan 2020
9. Heijungs, R. (1992). *Environmetal life cycle assessment of products – A guide of practice.* Centre pf Environmental Centre (CML).
10. Kristensen, T., & Kristensen, S. (2017). *Proportion, type and utilization of grassland affects the environmental impact of dairy farming.* http://labos.ulg.ac.be/dairyclim/de/literature/. Accessed 15 Jan 2020.
11. Lioy, R., Reding, R., Dusseldorf, T., & Meier, A. (2012). *CO2-emissions of 63 luxembourg livestock farms: A combined environmental and efficiency analysis approach.* EMILI-Congress (Emission of Gas and Dust from Livestock) – Saint-Malo, France – June 10-13, 2012. https://hal.archives-ouvertes.fr/hal-01190848/document. Accessed 15 Jan 2020.
12. Lioy, R., Dusseldorf, T., Meier, A., Reding, R., & Turmes, S. (2014). Carbon footprint and energy consumption of Luxembourgish dairy farms. 11. In *IFSA Symposium,* Berlin 1–4 April 2014. http://ifsa.boku.ac.at/cms/fileadmin/Proceeding2014/WS_2_7_Lioy.pdf. Accessed 15 Jan 2020.
13. Lioy, R., Meier, A., Dusseldorf T., Reding, R., & Thirifay, C. (2016). Sustainability assessment in Luxembourgish dairy production by CONVIS: A tool to improve both environmental and economical performance of dairy farms. In *The 12th IFSA Symposium 2016,* Harper Adams University, UK on 12–15 July 2016. http://ifsa.boku.ac.at/cms/fileadmin/Proceeding2014/WS_2_7_Lioy.pdf. Accessed 15 Jan 2020.
14. https://www.r-project.org/. Accessed 15 Jan 2020.
15. Dagnelie, P. (1994). *Théorie et methods statistiques* (Vol. 2, 8th ed.). Gembloux.
16. Bracher, A. (Ed.). (2011). *Möglichkeiten zur Reduktion von Ammoniakemissionen durch Förderungsmaßnahmen beim Rindvieh (Milchkuh).* SHL, Agroscope.
17. Seó, H., Pinheiro, M. F., Ruviario, C., & Léis, C. (2017, April). Avaliação do Ciclo de Vida na bovinocultura leiteira e as oportunidades ao Brasil. *Engenharia Sanitaria e Ambiental, 22.* https://doi.org/10.1590/s1413-41522016149096
18. Ross, S. A., Chagunda, M. G. G., Topp, C. F. E., & Ennos, R. A. (2012). *Effects of forage regime and cattle genotype on the global warming potential of dairy production systems.* EMILI-Congress (Emission of Gas and Dust from Livestock) – Saint-Malo, France – June 10–13, 2012. https://hal.archives-ouvertes.fr/hal-01190848/document. Accessed 15 Jan 2020.
19. Grignard, A., Hennart, S., Laillet, C., Oenema, J., & Stilmant, D. (2013). Bilan gaz à effet serre des ateliers laitiers des fermes pilotes Dayriman wallonnes selon la method GHG. In: *Acts of the symposium "20 ans Rencontres Recherches Ruminants",* Paris, France, December 4–5, 2013.

20. Hermansen, J. E., & Kristensen, T. (2011). Management options to reduce the carbon footprint of livestock products. *Animal Frontiers, 1*, 33–39.
21. Rotz, C. A., Montes, F., & Chianese, D. S. (2010). The carbon footprint of dairy production systems through partial life cycle assessment. *Journal of Dairy Science, 93*, 1266–1282.

Open Access This chapter is licensed under the terms of the Creative Commons Attribution 4.0 International License (http://creativecommons.org/licenses/by/4.0/), which permits use, sharing, adaptation, distribution and reproduction in any medium or format, as long as you give appropriate credit to the original author(s) and the source, provide a link to the Creative Commons license and indicate if changes were made.

The images or other third party material in this chapter are included in the chapter's Creative Commons license, unless indicated otherwise in a credit line to the material. If material is not included in the chapter's Creative Commons license and your intended use is not permitted by statutory regulation or exceeds the permitted use, you will need to obtain permission directly from the copyright holder.

Dynamic and Localized LCA Information Supports the Transition of Complex Systems to a More Sustainable Manner Such as Energy and Transport Systems

Florian Ansgar Jaeger, Cornelia Sonntag, Jörn Hartung, and Katrin Müller

Abstract The paper gives a snapshot of the potential of LCA (life cycle assessment) data-based optimizations in control systems. The environmental burden of existing infrastructure can be significantly reduced during use phase. Four Siemens' applications in different fields with different lead indicators show how LCA assessments can be adapted to fulfil the requirements of such applications. The applications are power and air quality management use cases in the field of eMobility, building management, industrial process control and traffic management. The main methodological challenge solved is the provision of the necessary temporal and special resolution, as well as forecasting of parameters for scheduling of processes.

1 Introduction

Life cycle assessment (LCA) methodology has become common to assess products and services and even found its way into strategy processes of planning infrastructure to convert our cities into sustainable urban areas [1]. Infrastructure has very long life cycles. Our time to cope with global warming is running up quickly, and there is little doubt that we need to speed up our climate actions as humanity. But to reduce emissions in markets with long life cycles, where inefficient assets can't quickly be replaced with sustainable ones, proves slow. We therefore propose to use LCAs of infrastructure during operation to improve the environmental performance of these infrastructures. To integrate environmental target functions into control systems and reducing or shifting consumption can increase environmental performance compared to conventional, solely monetarily or functionally optimized control algorithms. The goal is to make LCAs fit for control systems.

F. A. Jaeger (✉) · C. Sonntag · J. Hartung · K. Müller
Siemens Corporate Technology, Research in Energy and Electronics, Energy Systems,
Sustainable Life Cycle Management and Environmental Performance Management,
Berlin, Germany
e-mail: florian_ansgar.jaeger@siemens.com

© The Author(s) 2022
Z. S. Klos et al. (eds.), *Towards a Sustainable Future - Life Cycle Management*,
https://doi.org/10.1007/978-3-030-77127-0_6

2   Four Applications

The four case studies used as examples to show the potential of embedding environmental target functions into control systems are building, energy and transport cases:

- Smart charging for bus depots: A use case using flexibility in charging time of buses during their stay in bus depots in order to charge at times, where the grid mix has low average emissions. The analysis is part of the Mobility2Grid project and funded by the German Ministry for Education and Research.
- Smart cooling: A campus air conditioning system, which uses an ice storage in order to shift power consumption for cooling aggregates to times, where the grid mix has low average emissions. The analysis is part of the EnBA-M project and funded by the German Federal Ministry for Economic Affairs and Energy.
- Smart chemistry, methanol from steel mill gases: A case study using flexibility in power consumption, making an otherwise highly emitting process reduces GHG emissions. The analysis is part of the Carbon2Chem project and funded by the German Ministry for Education and Research.
- City Air Management: An online service operative in Nuremberg which is used to forecast events of high air pollution on a 5-day horizon at a street site measurement station. It simulates different interventions for this period to select them at times of maximum efficiency.

The methods described are a combination of conventional LCA, executed in LCA software and conventional control systems including forecasting algorithms and optimization algorithms. From a pure LCA prospective, they are based on comparative LCA, since the optimizer, no matter if machine or human operator, has to select between different scenarios. Not all assessments cover the full life cycle.

3   Smart Charging for Bus Depots

To guarantee operations of electric bus depots, charging infrastructure is slightly oversized in order to compensate for high demand events such as very cold or hot weather, delayed buses, maintenance and many other inconveniences. This necessary flexibility creates times at which buses are not charged and the grid connection is not fully utilized. This case study is an ex post analysis of the potential of this flexibility to reduce carbon emissions by charging at times, where the grid provides power of low CO_2e emissions. Three scenarios are analysed:

- Plug and charge: The buses are connected to the charger and start charging at full power, as soon as they are parked after returning to the depot and going through their daily routine.
- Cost-optimized: The buses are charged at max. Power during the period where the cost for power at the day-ahead market is the lowest without exceeding the grid connection.

- GHG-optimized: The buses are charged at max. Power during the period where the average GHG emission per kWh is the lowest without exceeding the grid connection.

The energy demand and schedules of the bus operation are based on real data from 140 Berlin diesel buses. Due to range restrictions, many buses are assumed to opportunity charge on the route. This increases the flexibility in depots.

3.1 Method

All three scenarios have the same hardware requirements, which is why only the power consumption during operations is part of the assessment. Only bidirectional charging or regulating the charging power based on battery wear would result in the necessity of expanding the system boundary to include the battery production and end of life. To calculate the optimum charging times based on an economic and a GHG target function, dynamic prices or emission functions for power are necessary. The spot market provides economic cost. Taxes and T&D (transmission and distribution) are not included. The dynamic country-based GHG emission factors per kWh are calculated on a time resolution of 15 min for Germany. T&D and upstream emissions are included. The grid mix is known for this time resolution, and each share of each energy carrier is multiplied with its respective energy carrier, as common for annual emission factor aggregates for countries too.

Combined with the bus schedules, dynamic emission factors feed into an optimizer, which defines at which time the buses are charged. The optimizer is set to optimize according to the target functions of the three scenarios stated above (view Fig. 1 Optimization Problem). The secondary constraints are the times the bus is available for charging, 100% state of charge when leaving the depot, the charging power and the grid connection limit of the depot. GHG emissions and cost for power are added up according to the resulting charging schedules of the three scenarios.

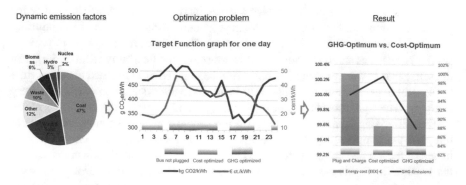

Fig. 1 Smart charging use case process

3.2 Results

The results show that the cost-optimized charging schedule does reduce the cost for power purchase at the energy market by almost 12% compared to a non-optimized plug and charge scenario. This results in a small increase in GHG emissions per kWh. The GHG-optimized charging schedule reduces GHG emissions by less than 0.5% and reduces cost by 4% compared to the plug and charge scenario (Fig. 1 on the right, GHG optimum vs. cost optimum).

3.3 Interpretation

The flexibility to shift charging times of buses is small. The shifting is only possible in the range of a few hours at maximum. The flexibility is almost exclusively available at night. There is almost no flexibility during daytime. But GHG emissions of the German grid mix don't frequently change drastically in short periods during the night, since there is no PV (photovoltaic power) at night and low-pressure zones for wind are moving slowly. This combination results in a marginal GHG saving potential of this application. In order to facilitate cost savings, however, the flexibility is relevant. Power prices at the power markets change more quickly at night, since the demand side has a larger impact. Cost-saving algorithms don't necessarily reduce GHG emissions as to be seen when looking at the results of the cost-optimized scenario.

4 Smart Cooling

Air conditionings are flexible loads. They are rarely running on full power, and any building has a certain thermal inertia, which can be used to store thermal energy. For this project, the thermal storage for the air conditioning was increased by adding a large ice storage to the system. The ice storage increases the temporal flexibility for power consumption. It can be charged independent of the demand of the building and discharged independent of the heat pump. This allows load shifting to provide similar services as smart charging. But in this case, the flexibility is much larger in the sense that power consumption can often be delayed or consumed ahead of time for several days. Additional complexity is added to the system compared to battery charging. The COP (coefficient of performance) and therefore the efficiency of the system differ significantly depending on the spread between the ambient temperature and the temperature of the thermal storage. These two parameters, plus the losses of the storage at high spreads over time, impact the overall power demand of the system. Since this system sets schedules in operation, the optimization is based on forecasted parameters for weather, power cost and emissions. The three scenarios and control mechanisms tested were similar to the smart charging case:

- Reference Scenario: System running without making use of the storage.
- Cost-optimized: The ice storage is filled at times with the best ratio of low power cost and high COP.
- GHG-optimized: The ice storage is filled at times with the best ratio of low relative GHG emissions for power and high COP.

As an additional indicator, the overall electricity demand is plotted.

4.1   Methodology

Even though the ice storage is not necessary for the operation according to the first scenario, production and end of life of the storage are not assessed. The storage was already available, but not in use, since cheap night rates for power had been abolished. The methodology of generating the environmental cost functions is the same as for bus charging above. But the data is based on forecasts for ambient temperatures, cooling demand of the buildings, cost and GHG emissions per kWh. The weather forecast is a commercially available API, and the other parameters are forecasted based on historical data of cooling demand and power generation mixes and day-ahead forecasts on renewable power generation and electricity load on the grid.

4.2   Results

Cost- and GHG-optimized operations are compared with the reference scenario. The cost-optimized operation shows a little reduction in power consumption of 0.2%, the GHG emissions increase by the same amount and the cost for power reduces by 4% (only cost at the power market). The GHG-optimized operation increases power consumption by 4% but reduces GHG emissions by 6%. Cost for power increases by almost 2% compared to the reference operation (Fig. 2: On the right).

4.3   Interpretation

It appears contradictory that the GHG-optimized operation leads to a higher power consumption. Figure 2 shows in the magnifier in the middle that the GHG-optimized operation leads to high power consumption in the middle of the day. This is due to the high availability of PV. The PV drives down the relative GHG emissions of the power mix at noon. This overcompensates the poor COP at daytime where ambient

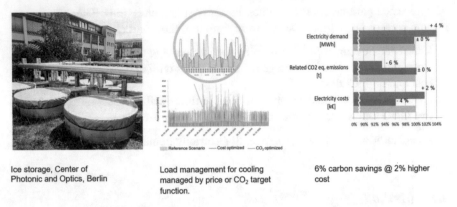

Ice storage, Center of
Photonic and Optics, Berlin

Load management for cooling
managed by price or CO_2 target
function.

6% carbon savings @ 2% higher
cost

Fig. 2 Smart cooling use case process

temperatures are high and the temperature spread between the ice storage and heat exchanger outside is high. Power cost is more demand driven; thus, cost can also be low at night, where the COP is more favourable. The case shows once again that cost- and GHG-optimized operations can lead to opposing results and create conflicts of interest.

5   Smart Chemistry, Methanol from Steel Mill Gases

The concept of carbon capture and use is to use CO_2 emissions from industrial processes and to reduce them with hydrogen in order to create basic chemicals such as methanol. This project uses electrolysis of water for the production of hydrogen. The target is to draw power when the load on the grid is lower than production in order to minimize curtailment of electricity from renewable energy sources. Using the fossil-based methanol production process as a benchmark, it was determined that, in addition, the power for electrolysis has to stay below 0.2 kg CO_2 eq./kWh with its GHG emissions to generate carbon savings.

5.1   Methodology

The methodology in use is very similar to the first two applications. Since it takes a long time to set up such a large-scale system, forecasting becomes inevitable even to determine the environmental performance of the first year of operation. A power scenario with an hourly resolution with times and volumes of excess energy is created in a multi-model scenario approach. It is based on publicly available plans and policies for the development of installed capacities of German power plants by energy carrier and the net structure in 2030, combined with appropriate

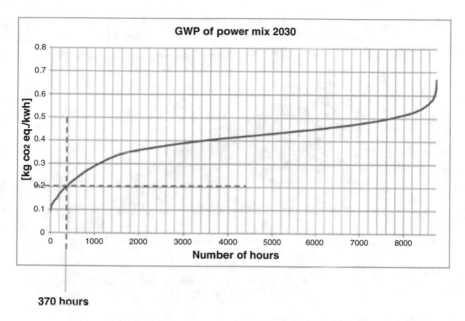

Fig. 3 GWP distribution of power

meteorological data. For the resulting power production profile, the greenhouse gas intensity of power production is calculated on an hourly base.

5.2 Results

For the underlying assumptions regarding the share of installed renewable energy sources, which results in a share of 47 per cent on gross power production, only 370 hours per year fulfilled the criteria of being below 0.2 kg CO_2 eq./kWh (Fig. 3). During these 370 h, the share of renewables in the power mix accounts for at least 70 per cent (Fig. 4). All these time periods coincide with periods of excess energy.

5.3 Interpretation

The analysis indicates that for the underlying assumptions on the share of renewables in power production, only few operating hours meet the criteria of low enough greenhouse gas emissions. A fluctuating electrolysis therefore would require immense capacities for electrolysis and hydrogen storage which cannot be implemented in practice due to economic reasons and required space. Moreover, hydrogen storage would lead to additional environmental impacts, not covered by this

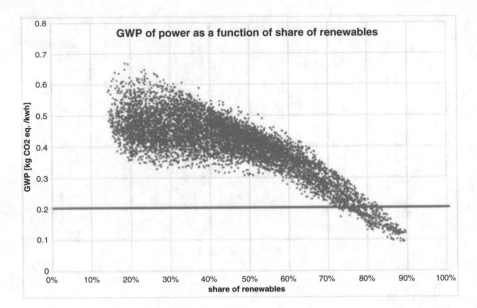

Fig. 4 Share of renewables at 200 g CO_2 eq./kWh

analysis. Hydrogen electrolysis during hours not meeting the low greenhouse gas level would cause a net increase of global warming impact of the CCU concept in comparison with the conventional processes of steel and methanol synthesis. This analysis is very sensitive to the assumed share of renewables and thus curtailment. Political targets for renewables have just been raised after the analysis. The potential of using excess energy for electrolysis will be recalculated under the new framework. The remaining hydrogen demand should be covered by hydrogen directly produced from renewable energy sources.

6 City Air Management

The City Air Management is an online web service which helps cities to manage local air quality at roadside measurement stations for the next 5 days (Fig. 5). It provides three basic functionalities for the air pollutants PM10, PM2.5 and NO_2:

- Monitoring the air quality at public measurement stations on a dashboard.
- Forecasting of air pollutants at these locations for 5 days.
- Intervention simulation and calculating pollution reduction of measures.

Instead of taking year-round measures, cities can take action when and where they have the highest impact.

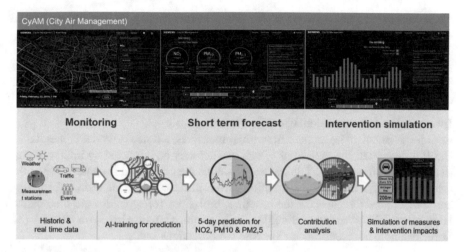

Fig. 5 City air management visualization and process

6.1 Methodology

Monitoring

Cities/counties/states operate their own air pollution sensor networks in order to prove their compliance with national or international regulation. This data is gathered on central servers and publicly available in most parts of the world. The CyAM has an API which allows this data to be pulled from this server or pushed to the CyAM as soon as the data is available. This is commonly every hour. The data for the individual measurement stations is visualized, categorized and benchmarked against the legal thresholds in a dashboard. It provides an immediate evaluation of the current situation and information on whether it is necessary to act. The latest history is also available for review, as well as the gliding annual average.

Forecasting

There are two common options to do forecasting for air pollutants, domain models and artificial intelligence. Domain models in this case are models which understand the physical and chemical processes of emission source behaviour and the atmospheric processes during transmission of pollutants. There are a vast variety of emission sources in and around a city. It involves tremendous efforts to assess all relevant fractions in real time. The modelling of the transmission (distribution plus the physical and chemical processes of the pollutants in the air) is time consuming, requires high computing capacities and is very sensitive to poor weather forecasts.

Thus, CyAM uses artificial intelligence to forecast air pollution concentrations at individual air quality measurement stations. It takes few available parameters which are available as forecasts. With historic data, it builds a temporal algorithm based on standard error backpropagation [2]. CyAM also uses air pollution measurement data, weather data/weather forecast data, calendric data and special events. The AI finds correlations and patterns in this data to predict air pollution for individual measurement stations. It doesn't contain any knowledge about the physical and chemical processes, responsible for these concentrations. Based on real-time data and forecasts of weather – and calendric/event data – a 5-day forecast is provided. The Advantage is a model which has high precision, takes little computing power during operation and requires few data points.

Intervention Impact Calculation
In order to calculate impacts of individual measures, a domain model is inevitable. But it only models the emissions which can actually be impacted by interventions, in this case traffic related. The traffic emissions are calculated for each hour of the following 5 days based on assumptions from historic data, calendric information and temperature forecasts for the baseline. Emissions for scenarios are then calculated for each intervention in SimaPro. Tailpipe emissions are based on HBEFA [3]. Some example interventions for specific street sections are:

- Allocation of eBuses on the lines passing the street section.
- Temporary driving ban of trucks or diesel cars for the street section.
- Low emission zones for the street section.
- Public transport ticket for air pollution season.

The local traffic-related share of the forecasted concentrations at the hotspot measurement station is determined correlating the forecasts of individual measurement stations in different locations. The combination of the traffic emission scenarios, the emission forecast and the traffic-related contribution of the forecasted concentration enables the prediction of the interventions' impact (Fig. 5).

6.2   Results

The accuracy of the forecast is measured by identifying how many of the 30% most polluted days were accurately predicted 5 days ahead of time. For NO_2 at the most polluted measurement station in Nuremberg, which is the lead indicator and location, this is 80%. Since it is an operational web service, the results are visualized on a dashboard as to be seen in the top three screenshots of Fig. 5. To evaluate the efficiency of the traffic interventions, the very same methodology is used as an ex post evaluation during the consulting phase of the project when the city selects which interventions they would like to have on the dashboard. The efficiency increase of temporary vs. all year-round measures is visualized in Fig. 6. The graph shows the results of the flexible truck ban for an individual road section in Nuremberg as a sum curve. The impact of the intervention is calculated for every day, relative to the annual saving. The jagged line is the historical sum curve from January 1 (on the very left, day 1) to December 31 (on the very right, day 365); see x-axis for the number of days per year. The smooth line sums up these savings, starting with the most efficient day of the year (on the very left, day 1), no matter if it is in January

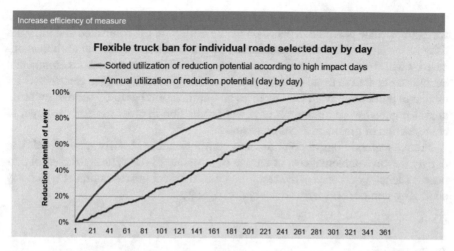

Fig. 6 Example efficiency of temporary vs. all year-round measures

or December, ending with the least efficient day of the year (on the very right, day 365).

6.3 Interpretation

The spread between these two lines shows the potential efficiency increase by implementing an intervention on a temporary basis, compared to an all year-round implementation. If a street section driving ban for trucks was implemented on the 70 most efficient days of the year, the yield in air pollution savings at the measurement station would be 50% of an all year-round implementation. In return, the least efficient 200 days, where there is enough wind to reduce emissions, only yield 20%. The efficiency of interventions measured as local air quality increase over days with traffic restrictions is most significant the fewer days they are triggered. Due to the fact that the forecast is not 100% accurate, the efficiency of the operational system is slightly lower, but cannot be determined at this stage of the project. Despite the efficiency, few cities apply such methods until now [4].

7   Conclusion

The use of environmental target functions in control systems has large potentials, reducing both global and local environmental impacts. Even when compared to economically optimized control strategies, environmental target function-based optimization can deliver significantly better environmental results. This is true even if cost of GHG emissions is priced in to some degree already at the energy markets, for example. The potential depends on the flexibility that is controlled and the volatility of the environmental impact. For some systems, environmental optimization-based control systems become absolutely crucial to create net environmental benefits compared to fossil-based processes. A large-scale hydrogen electrolysis for methanol production from CO_2 requires an optimization based on short-term prognosis for global warming impact of power production in order to meet the target of net reduction of greenhouse gas emissions.

From a methodological point of view, conventional LCA software and tools can deliver the environmental cost or burden of any state of the system for control purposes. The temporal and spatial resolution has to reflect the resolution at which any control system or short-term advisory tool operates.

References

1. Holst, J., Mueller, K., Jaeger, F. A., et al. (2018). *The city performance tool – How cities use LCM based decision support, designing sustainable technologies products and policies.* Springer.
2. Zimmermann, H. G., Tietz, C., & Grothmann, R. (2012). Forecasting with recurrent neural networks: 12 tricks. In G. Montavon, G. B. Orr, & K. R. Müller (Eds.), *Neural networks* (Vol. 7700). Springer.
3. Keller, M., Wuethrich, P., Ickert, L., Schmied, M., & Stutzer, B. et al. *Handbook emission factors for road transport*, 3.2 Ed., INFRAS AG, 25.7.2014.
4. Diegmann, V., Düring, I., Schönharting, J., et al. (2020). *Dynamisches umweltsensitives Verkehrsmanagement.* Verkehrstechnik Heft 321. https://bast.opus.hbz-nrw.de/opus45-bast/frontdoor/deliver/index/docId/2335/file/V321_barrierefrPDF.pdf. Accessed 18 Apr 2020.

Open Access This chapter is licensed under the terms of the Creative Commons Attribution 4.0 International License (http://creativecommons.org/licenses/by/4.0/), which permits use, sharing, adaptation, distribution and reproduction in any medium or format, as long as you give appropriate credit to the original author(s) and the source, provide a link to the Creative Commons license and indicate if changes were made.

The images or other third party material in this chapter are included in the chapter's Creative Commons license, unless indicated otherwise in a credit line to the material. If material is not included in the chapter's Creative Commons license and your intended use is not permitted by statutory regulation or exceeds the permitted use, you will need to obtain permission directly from the copyright holder.

Applying the Life Cycle Assessment (LCA) to Estimate the Environmental Impact of Selected Phases of a Production Process of Forming Bottles for Beverages

Patrycja Bałdowska-Witos, Robert Kasner, and Andrzej Tomporowski

Abstract The study concerns the current issues of the impact of packaging on the natural environment. The main goal was to analyse the life cycle (LC) of a beverage bottle made of polyethylene terephthalate. The functional unit comprised a total of 1000 PET bottles with a capacity of 1 l. The limit of the adopted system included steps from the moment of delivery of preforms to the production plant until they were properly shaped in the process of forming beverage bottles. Excluded from the system were the further stages of the production process, such as beverage bottling, labelling or storage/distribution. The processes related to the transport and storage of the raw material were also excluded. The LCA analysis was performed using the program of the Dutch company Pre Consultants called SimaPro 8.4.0. The "ReCiPe 2016" method was selected for the interpretation of lists of emitted chemicals. The results of the tests were presented graphically on bar charts and verified and interpreted.

1 Introduction

Activities of environmental organizations aimed at the development of pro-ecological behaviour of the population effectively communicate about positive and negative environmental impacts [1]. The model of behaviour shaped over the years has led to the development of various methods for identifying the occurrence of environmental threats [2]. An example of a method successfully implemented in industrial practice is the more and more frequently used life cycle assessment (LCA) [3, 4]. The LCA technique represents a new approach to assessing the potential environmental impacts of the beverage bottle manufacturing process. The growing ecological awareness of the society obliges production plants to carry out

P. Bałdowska-Witos (✉) · R. Kasner · A. Tomporowski
Department of Technical Systems Engineering, Faculty of Mechanical Engineering, University of Science and Technology in Bydgoszcz, Bydgoszcz, Poland
e-mail: patrycja.baldowska-witos@utp.edu.pl

© The Author(s) 2022

Z. S. Klos et al. (eds.), *Towards a Sustainable Future - Life Cycle Management*,
https://doi.org/10.1007/978-3-030-77127-0_7

environmental analysis. Such behaviour forces enterprises to strive for continuous improvement of the production process [4]. Change or modernization of technology should limit or minimize negative environmental impacts, take care of the environment and reduce or eliminate the negative effects of the production process [5]. The paper presents the results of the assessment of environmental impacts used in the technological process of bottle production. The goal of the study was to determine the potential levels of impact of individual technological operations on the condition of the natural environment and human health throughout the entire cycle of shaping bottles for beverages.

2   Materials and Methods

2.1   Research Methodology

The assessment of environmental loads was carried out for the production process of shaping bottles for beverages adopted in the study [6]. Collected industrial data from the bottle blow moulding machine made it possible to transform these data into the adopted functional unit. The analysis was performed using the ReCiPe 2016 method. Potential magnitudes of impacts from all environmental impacts were analysed [3, 6].

2.2   Determination of Goal and Scope

LCA is a tool used to assess the overall environmental impact of a product from "cradle to grave" [6]. For this purpose, the technological process of shaping PET bottles in Poland was assessed. The process is broken down into six unit operations, taking into account the demand for media and materials [6, 7]. The scope of the analysis included preform conveyor (CP), heating preforms (HP), stretching and lengthening the preform (SLP), blowing preforms (BP), degassing the bottle (DB) and cooling the finished bottle (CB).

2.3   System Boundary and Functional Unit

Six technological operations were adopted for the analysis (Fig. 1). As a result, the technological operations of the adopted processes were burdened with the same simplifications, which allowed assuming the exclusion level below 0.01% of the share in the entire life cycle. In inventory analysis, the examined systems and their system boundaries are defined, and process flow diagrams are drawn. Data collected

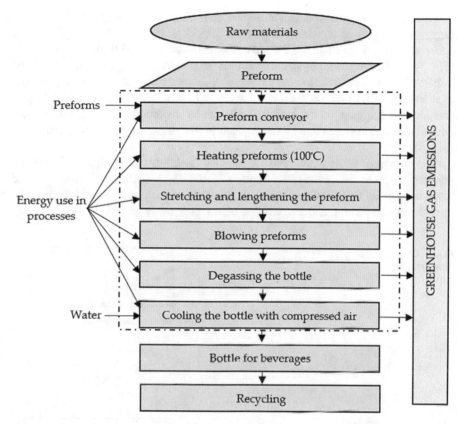

Fig. 1 Block diagram of the PET bottle production process

includes production, resource consumption and energy consumption. The functional unit adopted for the research was determined on the basis of data collected from the production company. It describes the production of 1000 bottles with a capacity of 1 l.

3 Results

The first stage of the research included defining the objectives and scope of the analysis, including checking the completeness and compliance of the adopted measurement data. The second stage of the research included the results of the analysis of the set of inputs and outputs. Developed on the basis of analytical results, it can be concluded that greenhouse gases responsible for the greenhouse effect, ultimately causing global warming, such as carbon dioxide and methane, are often released to the atmosphere from natural causes and anthropogenic origin [4, 8].

Potentially, the greatest negative impact on climate change was recorded for the degassing process of the finished product (DALY 1.16347E-08) (Fig. 2). Lower emission levels were observed for terrestrial ecosystems (3.51106E-11 species.yr) (Fig. 3) and freshwater ecosystems (9.5902E-16 species.yr) (Fig. 4). All of the three impact categories presented show the share of the raw material in the entire process of shaping the PET bottle at the level of approx. 78% of the impact in a given impact category. In the case of the stratospheric ozone depletion category, the highest potential environmental damage was caused by the degassing of the finished bottle (Fig. 5).

The emissions of non-carcinogenic toxicity for human were highest during the degassing step of the shaped PET bottle (8.04032E-11 DALY), while the second value in terms of emission was the bottle pressure forming process and the process of automatic stretching and lengthening of the previously heated preform (Fig. 7).

Fig. 2 Global warming, human health

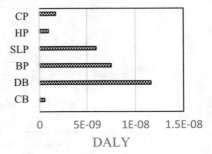

Fig. 3 Global warming, terrestrial ecosystems

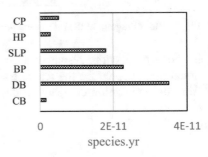

Fig. 4 Global warming, freshwater ecosystems

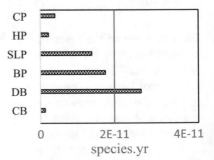

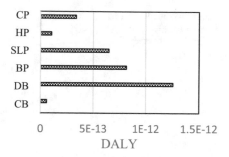

Fig. 5 Stratospheric ozone depletion

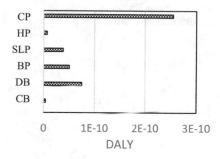

Fig. 6 Human carcinogenic toxicity

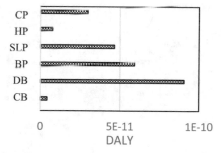

Fig. 7 Human non-carcinogenic toxicity

The total share of PET material in the technological process of shaping a PET bottle was only 1.76E-09 DALY. Significantly lower emission levels were observed for the whole human carcinogenic toxicity category (Fig. 6). The source of electricity is largely responsible for the amount of non-carcinogenic compounds emitted, and the amount of their emissions increases in stages as the production process progresses.

The ozone layer lies in the Earth's atmosphere and plays a key role in protecting living forms of nature [4]. Based on the analysis, it is proved that the process of shaping bottles exhibits greater environmental damage in the case of category zone formation, human health (Fig. 8) than in the case of category ozone formation, terrestrial ecosystems (Fig. 9).

Ecotoxicity of the aquatic and terrestrial environment results from the release of poisonous and toxic substances into the environment. Freshwater ecotoxicity shows the highest potential emission value specified for the degassing process of PET bottles (1.74682E-11 species.yr) (Fig. 11). The greatest potential impact on terrestrial ecotoxicity was exerted by degassing the PET bottle and was 1.79512E-11

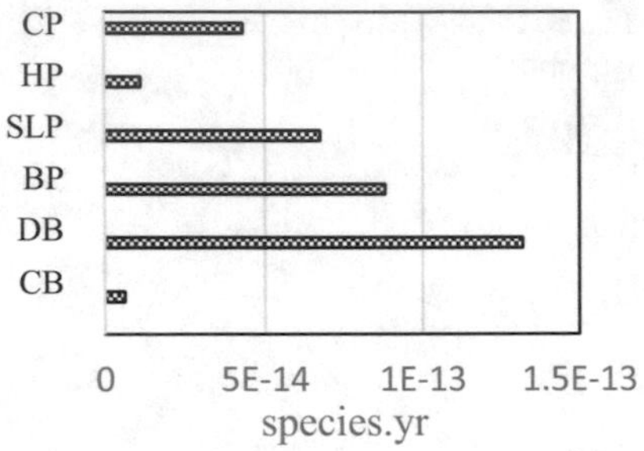

Fig. 8 Ozone formation, human health

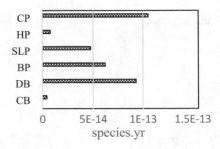

Fig. 9 Ozone formation, terrestrial ecosystems

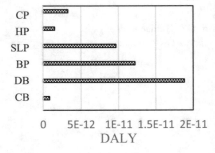

Fig. 10 Terrestrial ecotoxicity

species.yr. (Fig. 10). Among the six analysed unit processes, the lowest negative impact was noted for the cooling process of the shaped bottle.

Acidification of the terrestrial environment is caused by a lowering of the pH value. This phenomenon occurs as a result of disturbance of the ecological balance of the processes of energy and matter exchange between elements of ecosystems [4]. The process of cooling the finished product had the lowest negative impact in the entire shaping process, while the degassing process of the bottle had the greatest negative impact. Lower emission levels were observed for the terrestrial acidification category (Fig. 12). Characterizing the entire process of creating the bottle, the degassing process of the finished product (5.72E-12 species.yr) had the greatest impact on the land use category and slightly less (3.68E-12 species.yr) on the preform pressure shaping process, and nearly 1% of the impact was recorded for the preform-in-mould stretching and elongation process, and less than 1% for the preform processes prior to heating, heating and cooling the finished product (Fig. 13).

Ionizing radiation is a phenomenon that has always been present in the surrounding environment [4]. The greatest potential negative impact was emitted by one of

Fig. 11 Freshwater ecotoxicity

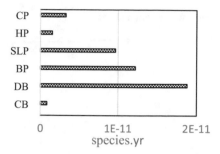

Fig. 12 Terrestrial acidification

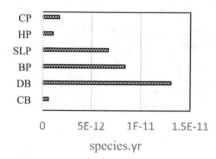

Fig. 13 Land use

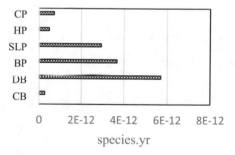

the sub-processes related to the amount of finished/semi-finished product used in the process – PET preform. The value of the issue was 2.11051E-13 DALY (Fig. 14). Local values of air pollutants generated by production plants do not provide accurate information on the scale of local emissions. As a result of the conducted analysis, it was determined that the greatest negative impact of the bottle shaping process on human health was recorded for the process of creating a PET bottle (1.18812E-08 DALY) at the stage of degassing the finished product. The lowest value of negative particle emissions affecting human health was determined for the PET bottle cooling process (4.59705E-10 DALY) (Fig. 15).

The PET bottle shaping process showed the higher potential level of adverse effects in the preform collection process for the reheating oven for the mineral resource scarcity category (Fig. 16) than for the fossil resource scarcity (Fig. 17). With the growing global demand for mineral resources, it is important to analyse whether the resources of geologically and technically available minerals in the Earth's crust can meet the future needs of humanity. Increasing recycling, material

Fig. 14 Ionizing radiation

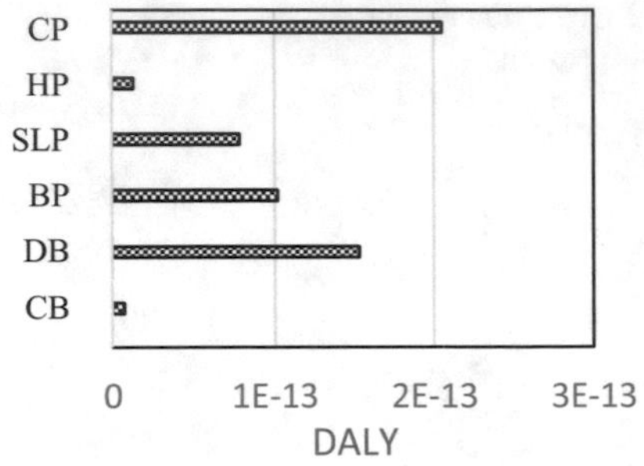

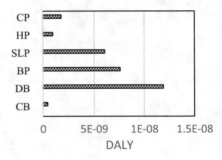

Fig. 15 Fine particulate matter formation

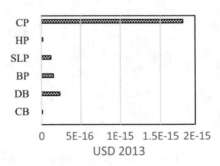

Fig. 16 Mineral resource scarcity

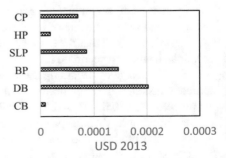

Fig. 17 Fossil resource scarcity

efficiency and demand management will surely play an important role in meeting future generations. A significant high value of potential impacts was recorded for the fossil resource scarcity category. The bottle with the greatest impact on this category was the PET bottle (8.58689E-05 USD2013) in the preform stretching and elongation process; the process of taking the preforms to the heating furnace (7.00854E-05 USD2013) was responsible for a slightly smaller amount of negative effects. This phenomenon is related to the fact that PET requires continuous extraction of fossil fuels, resulting in their depletion. This proceeding confirmed the highest staged impact of bottle production, and therefore the PET bottle probably had the greatest impact in the mineral extraction category. The production of the bottle had a slightly smaller impact in this category, possibly due to the fact that resource extraction is also needed.

Water consumption has an impact on human health and the quality of aquatic and terrestrial ecosystems. Water is critical to industry, the planet and people around the world. The production plant is a tycoon in the production of beverage bottles in the world. It uses a significant amount of the raw material, which is polyethylene terephthalate. However, its production significantly affects the condition of the natural environment. It was shown that for each of the two analysed impact categories, it was

Fig. 18 Water consumption, human health

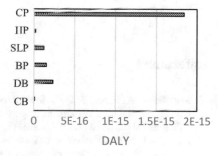

Fig. 19 Water consumption, aquatic ecosystems

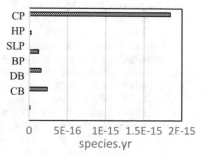

the preform conveyor operation that showed the highest potential negative environmental damage (Figs. 18 and 19).

4 Conclusion

The presented analyses of the technological processes of shaping bottles for beverages are characterized by various potential impacts on the condition of the natural environment. The environmental life cycle assessment [9] allowed for the conclusion that at the operational stage, the greenhouse gas emission index depends on the amount of electricity used in the production process [10]. The high level of impact on human health is determined by the raw material used in the production process – polyethylene terephthalate. In order to reduce the emission of negative impacts on the natural environment, producers of the food sector should constantly look for substitutes for raw materials from exhaustible fossil resources. The right direction of development is the popularization of biodegradable raw materials of natural origin. There is an urgent need for further research related to the environmental assessment of the beverage bottle manufacturing processes. In further considerations, the development of a method of waste management in the production process should be considered.

References

1. Mannheim, V., Fehér, Z. S., & Siménfalvi, Z. (2019). Innovative solutions for the building industry to improve sustainability performance with life cycle assessment modelling. In *Solutions for sustainable development* (pp. 245–253). CRC Press.
2. Piasecka, I., Bałdowska-Witos, P., Piotrowska, K., & Tomporowski, A. (2020). Eco-Energetical life cycle assessment of materials and components of photovoltaic power plant. *Energies, 13*, 6.
3. Bałdowska-Witos, P., Kruszelnicka, W., Kasner, R., Tomporowski, A., Flizikowski, J., Kłos, Z., Piotrowska, K., & Markowska, K. (2020). Application of LCA method for assessment of environmental impacts of a Polylactide (PLA) bottle shaping. *Polymers, 12*, 388. https://doi.org/10.3390/polym12020388
4. Bałdowska-Witos, P., Kruszelnicka, W., Kasner, R., Rudnicki, J., Tomporowski, A., & Flizikowski, J. (2019). Impact of the plastic bottle production on the natural environment. Part 1. Application of the ReCiPe 2016 assessment method to identify environmental problems. *Przemysl Chemiczny, 98*(10), 1662–1667.
5. Kłos, Z. (2002). Ecobalancial assessment of chosen packaging processes in food industry. *International Journal of Life Cycle Assessment, 7*, 309.
6. Bałdowska-Witos, P., Kruszelnicka, W., Kasner, R., Tomporowski, A., Flizikowski, J., & Mrozinski, A. (2019). Impact of the plastic bottle production on the natural environment. Part 2. Analysis of data uncertainty in the assessment of the life cycle of plastic beverage bottles using the Monte Carlo technique. *Przemysl Chemiczny, 98*, 1668–1672.
7. Kruszelnicka, W., Marczuk, A., Kasner, R., Bałdowska-Witos, P., Piotrowska, K., Flizikowski, J., & Tomporowski, A. (2020). Mechanical and processing properties of Rice grains. *Sustainability, 12*, 552. https://doi.org/10.3390/su12020552

8. Bałdowska-Witos, P., Kruszelnicka, W., Kasner, R., Tomporowski, A., Flizikowski, J., Kłos, Z., Piotrowska, K., & Markowska, K. (2020). Application of LCA method for assessment of environmental impacts of a Polylactide (PLA) bottle shaping. *Polymers, 12*, 388. https://doi.org/10.3390/polym12020388
9. Piasecka, I., Bałdowska-Witos, P., Piotrowska, K., & Tomporowski, A. (2020). Eco-Energetical life cycle assessment of materials and components of photovoltaic power plant. *Energies, 13*, 1385. https://doi.org/10.3390/en13061385
10. Joachimiak-Lechman, K., Selech, J., & Kasprzak, J. (2018). Eco-efficiency analysis of an innovative packaging production: Case study. *Clean Technologies and Environmental Policy, 21*(2), 339–350. https://doi.org/10.1007/s10098-018-1639-7

Open Access This chapter is licensed under the terms of the Creative Commons Attribution 4.0 International License (http://creativecommons.org/licenses/by/4.0/), which permits use, sharing, adaptation, distribution and reproduction in any medium or format, as long as you give appropriate credit to the original author(s) and the source, provide a link to the Creative Commons license and indicate if changes were made.

The images or other third party material in this chapter are included in the chapter's Creative Commons license, unless indicated otherwise in a credit line to the material. If material is not included in the chapter's Creative Commons license and your intended use is not permitted by statutory regulation or exceeds the permitted use, you will need to obtain permission directly from the copyright holder.

Part II
Sustainable Technologies

Accounting for the Temporal Fluctuation of Wind Power Production When Assessing Their Environmental Impacts with LCA: *Combining Wind Power with Power-to-Gas in Denmark*

Romain Besseau, Milien Dhorne, Paula Pérez-López, and Isabelle Blanc

Abstract Worldwide wind power capacity is increasing, while the environmental footprint and economic cost of energy produced decrease. However, wind power generation is weather-dependent. At a high penetration rate, storage systems such as power-to-gas may become necessary to adjust electricity production to consumption. This research work presents the environmental life cycle performance of wind power accounting for the energy storage induced by the temporal variability of weather-dependent production and consumption. A case study in which wind power installations are combined with a power-to-gas system in Denmark to provide electricity according to the national load consumption profile was considered. Results highlight an increase, roughly by a factor 2, of the carbon footprint coming from both energy storage infrastructure and induced losses, but remain significantly, at least ten times, lower than fossil counterparts.

1 Introduction

Renewable energy systems (RES), promoted to limit the dependency of the energy mix to fossil fuel, and the environmental impact associated with their use are currently prospering at a global level [1].

Although RES are based on the exploitation of renewable sources, this does not mean that the renewable energy generated is impact-free. Indeed, energy and materials are necessary to build, operate, and dismantle those systems. Life cycle assessment (LCA) is an appropriate tool, often applied to assess the environmental footprint of RES [2]. LCA results published in the literature highlight that RES generally present significantly lower environmental footprint over the life cycle than fossil fuel-based alternatives [3].

R. Besseau (✉) · M. Dhorne · P. Pérez-López · I. Blanc
Mines ParisTech, PSL University, Paris, France
e-mail: romain.besseau@mines-paristech.fr

© The Author(s) 2022
Z. S. Klos et al. (eds.), *Towards a Sustainable Future - Life Cycle Management*,
https://doi.org/10.1007/978-3-030-77127-0_8

Moreover, it is positive to note that along with the development of RES industries, the efficiency of systems and the underlying manufacturing processes have improved, leading to better environmental performance as well as economic performance with time [4, 5]. These improvements pave the ways to a massive deployment of affordable and low environmental footprint renewable energy.

However, the electricity production of RES can be weather-dependent, as in the case of wind power, and not necessarily in adequacy with consumption. As a consequence, the massive integration of these technologies into the electricity mix requires the use of either dispatchable power plants or storage systems to be able to balance production with the consumption load profile at any time and thus maintain the grid stability [6].

With a wind power production equivalent to 45% of the annual electricity consumption in 2017, Denmark is an example of a country with high penetration of RES [7]. Thus, the balance between production and consumption strongly relies on the interconnection with Baltic countries and local combined heat and power (CHP) plants. Baltic countries are richly endowed with hydropower plants that can adjust their own production and even pump back water to store energy [8]. Denmark is also equipped with particularly flexible CHP plants [9]. Investments have been done to lower their minimum power output, provide overload ability, increase their ramping speed, and reduce the cost and time to stop and start power generation.

Existing hydropower capacities are limited, and the potential for new hydropower installations equipped with large reservoirs remains low in Europe [10]. In addition, the use of CHP plants relying on fossil fuel must be reduced as low as possible to mitigate climate change. As a consequence, new solutions for further integration of weather-dependent RES become necessary.

One of those solutions is the use of power-to-gas (P2G) systems, which consist in turning electric power into synthetic gas. Electricity is used to hydrolyze water molecules and generate dihydrogen (H_2). This gas can be stored and used directly or turned into methane (CH_4) after an additional transformation called methanation. P2G, by coupling electric and gas grids, offers the possibility to store massive amount of energy over long periods of time. Thus, P2G is a storage technology able to provide long-term potentially seasonally or annually contrary to electrochemical batteries that are limited to short-term storage [11]. Once stored, the gas can be used to generate back electricity in a gas power plant or be used for mobility purposes, heat, or industrial uses. For those reasons, IEA [12] and other energy experts [11] see P2G as a determinant technology for electric systems' operation.

As RES themselves, storage technologies require materials and energy to be manufactured, operated, and dismantled and therefore involve environmental burdens. Few LCA studies have been published and most of them focus on mobility applications [13, 14]. For such applications, the P2G system is continuously used to maximize gas production and does not adapt to the fluctuations of RES production. As a consequence, P2G systems considered for mobility applications present ultimate load factors of 91% approximately, corresponding to 8000 h/year at full load [14]. This level is much higher than what would correspond to a P2G system designed to cope with the variability of renewable energy sources. Thus, LCA results calculated for mobility applications cannot be directly extrapolated to assess the environmental performance of P2G systems designed to balance the fluctuation of renewable energy production.

Consequently, we assessed the environmental life cycle performance of renewable energy accounting for the energy storage induced by the temporal variability of weather-dependent production and consumption. A case study in which wind power installations are combined with a power-to-gas system in Denmark to provide electricity according to the Danish load consumption profile was considered. Denmark has been chosen as wind power is highly developed with, in 2017, a production corresponding to 45% of electricity consumption [7], which is expected to increase [7], and P2G technologies already under study with a project of P2G demonstrator [15].

2   Material and Methods

To assess the environmental performance of a system composed of wind turbines combined with P2G storage, the following elements need to be modeled and quantified:

1. The environmental impacts of wind turbines.
2. The environmental impacts of the components of a P2G system.
3. The need and use of storage.

Environmental impacts were calculated using the Python library *Brightway2* [16] dedicated to LCA and using the cutoff version of ecoinvent 3.4 for background life cycle inventories.

2.1   Environmental Impacts of the Wind Turbine

Environmental impacts of wind turbines are assessed making use of the parametric LCA model developed and presented in detail in [5, 17]. This parametric model uses the LCA-specific Python library *Brightway2* and can be accessed online at https://github.com/romainsacchi/LCA_WIND_DK. It enables to create tailor-made life cycle inventories of wind turbines considering their specific technological characteristics and fitting their spatiotemporal context.

Onshore and offshore wind turbines have been selected to have an environmental performance representative of the fleet. Their nominal power is 3.6 MW, and rotor diameter is 120 m corresponding to a power density ratio of 310 W/m^2. The onshore wind turbine has a 95 m hub height, while the offshore turbine has an 85 m hub height. The offshore turbines are considered to be grouped in a farm of 50 wind turbines located 5 km from shore, with a sea depth of 5 m. Onshore and offshore wind turbines are exposed to wind resource leading to load factors of respectively 30% and 50% in coherence with measured production in Denmark [5].

2.2 Environmental Impacts of the P2G Systems

As for assessing wind turbines' environmental impacts, a parametric model has been developed to assess the environmental impacts of the components of a P2G system.

A P2G system is composed of an electrolyzer, a methanation reactor requiring a prior system to capture CO_2 in case of P2M but not for P2H, and a power plant to generate electricity from the produced gas.

The electrolyzer is composed of cell stacks where electrolysis takes place, power electronics to feed the cells with the right current and voltage, and additional equipment such as pipes and reservoirs [18]. The cell stack is modeled by adapting the ecoinvent LCI of solid oxide fuel cell to represent the use of alkaline instead of solid oxide cell stack. To do so, lanthanum oxide is replaced by nickel and zirconium oxide by potassium. The power electronics is modeled using the existing inverter LCI originally created for photovoltaic systems. The additional equipment, which mainly consists of pipes and reservoirs, are modeled by the ecoinvent stainless steel pipe dataset. The methanation reactor is also modeled with stainless steel in accordance with the previous work from Zhang et al. [14]. Carbon dioxide is required for methanation reaction, so its extraction from the flue gas of an industrial chimney is modeled using inventories from Koornneef et al. [19]. Finally, the power plant used to burn the synthesized gas and generate electricity is modeled using the ecoinvent combined cycle gas turbine substituting the fossil gas by the synthesized gas.

The weight, lifetime, and efficiency of all the devices are based on data from industrial reports and scientific literature [18, 20].

2.3 Assessment of the Need and Use of Storage

The need and the use of storage are assessed from the comparison of energy production time series and consumption time series. The approach is represented in Fig. 1.

Firstly, wind power production time series are determined from wind speed time series and the wind turbine power curve. Wind speed data can come from on-site measurements or weather reanalysis data. MERRA-2 wind speeds have been downloaded from the online platform Renewables.ninja (https://www.renewables.ninja/) and are used in this study. Power curve gives the relationship between the wind speed the turbine is exposed to and the corresponding power output. Manufacturer power curves or modeled power curves can be used. A model able to generate a wind turbine power curve based on the nominal power, the rotor dimension and the wind turbulence intensity is used, as well as a wake loss coefficient when wind

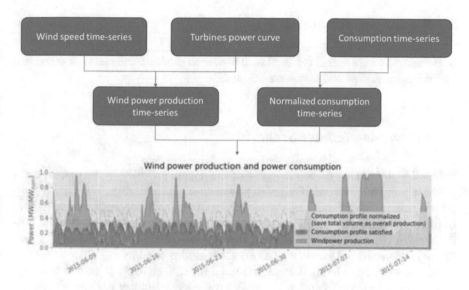

Fig. 1 Graphical representation of the approach used to assess the need and use of storage

turbines are grouped into wind farms [5]. The convolution of wind speed data with the power curve gives wind power production time series.

In a second step, this production time series is compared to the normalized Danish historical load curve. The blue curve (Fig. 1) represents the wind power production per MW installed. The orange dotted curve represents the Danish load curve with an annual consumption equivalent to the wind power production. When the production exceeds the consumption, the excess energy has to be stored, and when the production is lower, the energy difference has to be retrieved from storage. As storage induces energy losses, the amount of energy that can be retrieved from storage is lower than the amount of energy that is stored. As a consequence, a load curve corresponding to a lower annual consumption than the annual production can only be satisfied. The load curve that can be satisfied is calculated by considering the volume of energy that has to be stored and retrieved from storage and the storage efficiency.

Once the load curve is established, it is possible to get, as represented in Fig. 1:

- The wind power production in blue.
- The consumption that can be satisfied in red.
- The intersection of blue and red curves that gives the wind power production directly consumed.
- The difference between wind power production and the intersection of blue and red curves that gives the amount of energy that has to be stored when blue curve exceeds the red one, and retrieved from storage when the consumption exceeds the wind power production.

3   Results and Discussion

One scenario where energy is stored with power-to-methane and one with power-to-hydrogen are studied and discussed below.

3.1   Power-to-Methane Scenario

Figure 2A presents the carbon footprint of the energy provided by the system composed of wind turbines combined with power-to-methane storage. The carbon footprint is respectively 30 g CO_2eq/kWh and 20 g CO_2eq/kWh for onshore and offshore turbines. When neglecting the constraint related to weather dependency of the production, and thus the induced need for storage, the carbon footprint of energy produced is respectively 15 and 10 g CO_2eq/kWh as illustrated by Fig. 2B.

In that case, considering the induced need for storage leads to an increase by a factor 2 of the carbon footprint. Figure 2C presents the impact per power capacity installed and highlights an increase of the carbon footprint of the system wind turbines combined with P2M storage compared to wind turbines alone. However, the difference cannot be explained only by the addition of the storage infrastructure. The second reason leading to that increase of the carbon footprint is the storage energy loss. Figure 2D shows that around half of the wind power production is

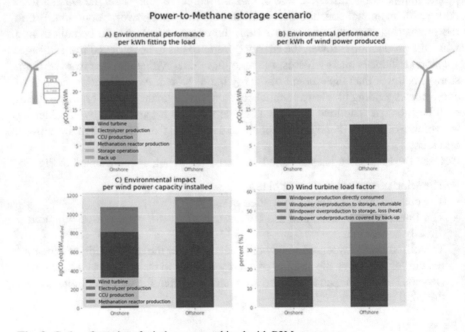

Fig. 2 Carbon footprint of wind power combined with P2M storage

directly consumed and half goes to the storage. Considering the energy going to the storage, 70% is dissipated as heat "loss" and 30% will be restituted as electricity. As a consequence, a significant part of the energy generated will be wasted as heat leading to an increase of the impact per kWh of electricity delivered by the system.

3.2 Power-to-Hydrogen Scenario

If H_2 can be stored over long periods, P2H could be used as an alternative to P2M and provides as well seasonal storage. Figure 3 presents the results considering a power-to-hydrogen scenario instead of power-to-methane. The carbon footprint of wind power combined with P2H is slightly higher than 25 g CO_2eq/kWh for onshore turbines and slightly lower than 20 g CO_2eq/kWh for offshore turbines. These values are lower than those of the power-to-methane scenario due to:

– A reduced storage infrastructure. In the absence of methanation reaction, there is no need for methanation reactors, no need for carbon capture.
– Reduced storage energy losses: the efficiency of the exothermic methanation reaction is limited to 74% [20]. Removing this step limits the decrease of the storage efficiency.

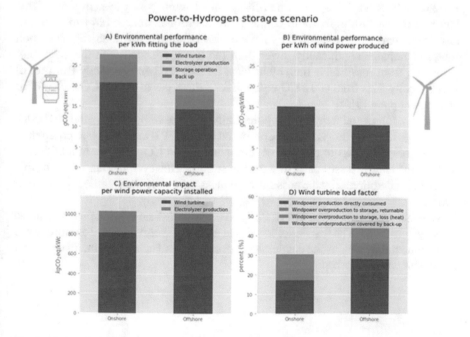

Fig. 3 Carbon footprint of wind power combined with P2H storage

However, the feasibility of this scenario is conditioned to feasibility of storing H_2 over long periods. Storing massive amounts of H_2 over long periods of time may be more complex due to its volatility; H_2 is the smallest molecule on earth. In some countries such as the Netherlands, it is already possible to inject up to 13% of H_2 into the national gas grid [21]. The feasibility of higher levels remains, to our knowledge, to be demonstrated.

Thus, P2H is an option that can be considered in association with wind power to limit the carbon footprint of energy delivered. If hydrogen storage is too complex for long-term storage, a system combining P2H for short-term storage with a higher efficiency and P2M to balance only long-term production variation can be contemplated.

4 Conclusion

An approach to account for the temporal fluctuation of wind power production when assessing their environmental impacts with LCA has been developed. This method has been applied to a case study in Denmark where wind power is combined with power-to-gas to deliver electricity according to the national load curve.

Wind power combined with P2G can deliver electricity as a dispatchable power plant with a low carbon footprint. The carbon footprint of the system "wind power + P2M" is around twice the carbon footprint of wind power system alone. This increase of the carbon footprint comes from additional storage infrastructure but also from energy losses induced by the storage. Despite being doubled, the carbon footprint remains significantly lower than the one associated with the electricity generated from fossil fuels (i.e., by at least a factor of 10). Electricity from fossil fuel typically goes from 400 g CO_2eq/kWh for natural gas to 1000 g CO_2eq/kWh for coal power plant [3].

The environmental footprint can potentially be reduced by limiting the power-to-gas storage to the hydrogen stage. Compared to P2M, P2H requires less equipment for the storage, thanks to the lack of methanation and CO_2 capture. In addition, it reduces the energy losses occurring during storage. However, the feasibility of this scenario is conditioned by the possibility to massively store the volatile H_2 over long periods.

Whether combined with P2M or P2H, an important share of wind power remains directly consumed. The rest is stored with significant energy losses. To improve the environmental as well as the economic performance of such a system, a key aspect is the heat waste valorization.

References

1. IRENA. (2017). *Renewable power generation costs in 2017* (p. 160). IRENA.
2. Asdrubali, F., Baldinelli, G., D'Alessandro, F., & Scrucca, F. (2015). Life cycle assessment of electricity production from renewable energies: Review and results harmonization. *Renewable and Sustainable Energy Reviews, 42*, 1113–1122. https://doi.org/10.1016/j.rser.2014.10.082
3. IPCC. (2012). Renewable energy sources and climate change mitigation: Special report of the intergovernmental panel on climate change. *Choice Reviews Online, 49*(11), 49-6309. https://doi.org/10.5860/CHOICE.49-6309
4. Louwen, A., van Sark, W. G. J. H. M., Faaij, A. P. C., & Schropp, R. E. I. (2016). Re-assessment of net energy production and greenhouse gas emissions avoidance after 40 years of photovoltaics development. *Nature Communications, 7*, 13728. https://doi.org/10.1038/ncomms13728
5. Besseau, R. (2019). Past, present and future environmental footprint of the Danish wind turbine fleet with LCA_WIND_DK, an online interactive platform. *Renewable and Sustainable Energy Reviews, 15*.
6. Seck, G. S., Krakowski, V., Assoumou, E., Maïzi, N., & Mazauric, V. (2017). Reliability-constrained scenarios with increasing shares of renewables for the French power sector in 2050. *Energy Procedia, 142*, 3041–3048. https://doi.org/10.1016/j.egypro.2017.12.442
7. Energinet. "Data: Oversigt over energisektoren," *Energistyrelsen*, 25-Aug-2016. [Online]. Available: https://ens.dk/service/statistik-data-noegletal-og-kort/data-oversigt-over-energisektoren. Accessed 17 Apr 2019.
8. Child, M., Bogdanov, D., & Breyer, C. (2018). The Baltic Sea Region: Storage, grid exchange and flexible electricity generation for the transition to a 100% renewable energy system. *Energy Procedia, 155*, 390–402. https://doi.org/10.1016/j.egypro.2018.11.039
9. Danish Energy Agency, Energinet, EA, CNREC, and Electric Power Planning & Engineering Institute. (2018). *Thermal power plant flexibility, a publication under the clean ministerial campaign.* Clean Energy Ministerial.
10. Gimeno-Gutiérrez, M., & Lacal-Arántegui, R. (2013). *Assessment of the European potential for pumped hydropower energy storage – A GIS-based assessment of pumped hydropower storage potential* (p. 74). Elsevier.
11. Blanco, H., & Faaij, A. (2018). A review at the role of storage in energy systems with a focus on power to gas and long-term storage. *Renewable and Sustainable Energy Reviews, 81*, 1049–1086. https://doi.org/10.1016/j.rser.2017.07.062
12. IEA. (2014). *Technology roadmap energy storage* (p. 64). IEA.
13. Wettstein. (2018). *LCA of renewable methane for transport and mobility* (p. 40). ZHAW.
14. Zhang, X., Bauer, C., Mutel, C. L., & Volkart, K. (2017). Life cycle assessment of power-to-gas: Approaches, system variations and their environmental implications. *Applied Energy, 190*, 326–338. https://doi.org/10.1016/j.apenergy.2016.12.098
15. "P2G-BioCat." [Online]. Available: https://biocat-project.com/. Accessed 17 Jan 2020.
16. Mutel, C. (2017). Brightway: An open source framework for life cycle assessment. *The Journal of Open Source Software, 2*(12), 236. https://doi.org/10.21105/joss.00236
17. Sacchi, R., Besseau, R., Pérez-López, P., & Blanc, I. (2019). Exploring technologically, temporally and geographically-sensitive life cycle inventories for wind turbines: A parameterized model for Denmark. *Renewable Energy, 132*, 1238–1250. https://doi.org/10.1016/j.renene.2018.09.020
18. Hydrogenics, "Electrolyzer," 2019. [Online]. Available: https://www.hydrogenics.com/wp-content/uploads/2-1-1-industrial-brochure_english.pdf?sfvrsn=2. Accessed 10 Oct 2019.
19. Koornneef, J., van Keulen, T., Faaij, A., & Turkenburg, W. (2008). Life cycle assessment of a pulverized coal power plant with post-combustion capture, transport and storage of CO2. *International Journal of Greenhouse Gas Control, 2*(4), 448–467. https://doi.org/10.1016/j.ijggc.2008.06.008

20. Electrochaea, "Data-Sheet BioCat Plant," 2019. [Online]. Available: http://www.electro-chaea.com/wp-content/uploads/2018/03/201803_Data-Sheet_BioCat-Plant.pdf. Accessed 10 Oct 2019.
21. Quarton, C. J., & Samsatli, S. (2018). Power-to-gas for injection into the gas grid: What can we learn from real-life projects, economic assessments and systems modelling? *Renewable and Sustainable Energy Reviews, 98*, 302–316. https://doi.org/10.1016/j.rser.2018.09.007

Open Access This chapter is licensed under the terms of the Creative Commons Attribution 4.0 International License (http://creativecommons.org/licenses/by/4.0/), which permits use, sharing, adaptation, distribution and reproduction in any medium or format, as long as you give appropriate credit to the original author(s) and the source, provide a link to the Creative Commons license and indicate if changes were made.

The images or other third party material in this chapter are included in the chapter's Creative Commons license, unless indicated otherwise in a credit line to the material. If material is not included in the chapter's Creative Commons license and your intended use is not permitted by statutory regulation or exceeds the permitted use, you will need to obtain permission directly from the copyright holder.

Integrated Life Cycle Sustainability Assessment: Hydrogen Production as a Showcase for an Emerging Methodology

Christina Wulf, Petra Zapp, Andrea Schreiber, and Wilhelm Kuckshinrichs

Abstract Ideally, life cycle sustainability assessment (LCSA) consists of life cycle assessment (LCA), life cycle costing (LCC) and social life cycle assessment (S-LCA) based on a joint technical model. For an integrated and consistent LCSA, however, this is not enough. Therefore, in this work, a coherent indicator selection based on the Sustainable Development Goals (SDGs) as well as an integration of the impact categories/indicators with the help of multi-criteria decision analysis is conducted. The chosen method PROMETHEE does not allow full compensation of the sustainability indicators, which reflects a possible view on sustainability. The SDG-based approach is compared with a classical approach where the weighting is based on the three sustainability dimensions. Both are tested on comparison case study of a 6 MW pressurized electrolyser located in three European countries, i.e. Spain, Germany and Austria, to illustrate the difference of industrial hydrogen production in industrialized countries with different structures of electricity markets.

1 Introduction

The Sustainable Development Goals (SDGs) published in 2015 by the UN [1] gain more and more importance. This is true not only for countries and for regions, for which they were drafted in the first place, but also for companies and academia. For life cycle sustainability assessment (LCSA), there are several approaches to link those two concepts. For example, the project "Linking the UN SDGs to life cycle impact pathway frameworks" [2] by 2.-0 LCA consultants and PRé consultants under the umbrella of the UN Life Cycle Initiative develops impact pathways for the SDGs, which are cause-effect oriented. These should, for example, serve as impact categories for the social life cycle assessment (S-LCA) [3]. Owsianiak et al. [4]

C. Wulf (✉) · P. Zapp · A. Schreiber · W. Kuckshinrichs
Forschungszentrum Jülich, Institute of Energy and Climate Research – Systems Analysis and Technology Evaluation, Jülich, Germany
e-mail: c.wulf@fz-juelich.de

© The Author(s) 2022
Z. S. Klos et al. (eds.), *Towards a Sustainable Future - Life Cycle Management*,
https://doi.org/10.1007/978-3-030-77127-0_9

have taken the SDG indicators related to the environment and have tested if they actually help to reach environmental sustainability. For that, they did not only take principles of life cycle assessment (LCA) into account but also the planetary boundaries. A rough match between the SDGs and LCSA indicators has been done by the authors in an earlier study Wulf et al. [5]. They assigned the often used LSCA indicators to the SDGs as well as their indicators.

These approaches concentrate mainly on indicator selection and impact assessment. A further topic of LCSA is the integration of indicators with the help of multi-criteria decision analysis (MCDA) [6, 7]. In this paper, it is presented how the SDGs can guide MCDA for LCSA. The implications of this approach in contrast to the understanding of sustainability based on three dimensions are discussed afterwards. The different effects are analysed on the example of comparing different locations for hydrogen production as an actual LCSA case study.

2   Methodology

In this paper, an LCSA is performed with the guidance from the SDGs (Fig. 1). They are used for the indicator selection as well as for the MCDA. The quantification of the different indicators is done with classical LCA, life cycle costing (LCC) and S-LCA, the latter performing a hot spot analysis. These SDG-guided indicator values describe the performance of the considered systems and form the input for the MCDA method PROMETHEE (Preference Ranking Organization METHod for Enrichment of Evaluations). In many studies, the three assessment methods are regarded as equally important because they are loosely representing the three dimensions of sustainability [6]. This premise is used to derive weighting factors for the different indicators. In this study, however, not the three dimensions of sustainability are considered as equally important, but each sustainability goal has the same importance. This leads to a different set of weighing factors than is the case with the three dimensions of sustainability. In this section, the relation between the SDGs and the LCSA indicators is explained in more detail as well as the choice of method for MCDA and how the weighting factor set is calculated.

Fig. 1 Approach for integrating the SDGs into LCSA

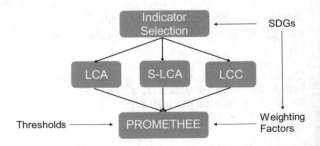

2.1 Sustainable Development Goals and LCSA Indicators

The assignment of LCSA indicators to the SDGs is based on the previous paper [5]. The indicators are selected based on common guidelines. For the LCA, these are the recommendations from the ILCD [8] and guidance documents by theUNEP/SETAC [9]. Indicators on the midpoint level are used as implemented in the GaBi software. The S-LCA indicators are based on the respective UNEP/SETAC guidelines [10] and their interpretation in PSILCA 2 [11] for a hotspot analysis. Indicators in PSILCA 2 tackling issues that are also assessed by LCA are excluded from the selection. The LCC indicators are guided by the European Investment Bank [12]. Particular attention has been paid to avoid double or triple counting of topics. The findings of this matching can be seen in Fig. 2.

It must be mentioned that goals 2, 11 and 17 cannot be described by LCSA indicators.

2.2 Multi-criteria Decision Analysis

When performing a full LCSA, a bundle of very different indicator values with physical, monetary and other units result. In such a case, MCDA can help to structure the decision-making process. Within this MCDA guidance process, fundamental value-based choices have to be made. In particular, it has to be decided to what extent compensation between indicators is allowed. In this work, compensation is not allowed. With respect to a value-based approach, this is a very crucial assumption. However, it helps to clarify the problem. As a specific method representing this, PROMETHEE II [14] is chosen. This method is based on a pairwise comparison of the different options. The most preferable option has the highest result, which is called outranking flow Φnet. A linear preference function with indicator-specific thresholds is applied [15].

2.3 Equal Weighting of SDGs

The premise of the indicator weighting of this paper is that each SDG has the same weight. Furthermore, indicators describing one SDG have the same importance. However, this results in unequal weighting of indicators in case of different numbers of indicators per SDG. Additionally, there are some indicators describing not only one goal but two or more. For example, trade union (density as a % of paid employment total) is describing goal 8 (decent work and economic growth) as well as goal 16 (peace, justice and strong institutions) (see Fig. 1). To avoid an overestimation of such indicators, the number of indicators in one goal m is normalized with the number of assigned goals p. This is mathematically expressed in Eq. 1.

Fig. 2 SDGs and their respective LCSA indicators, icons from [13]; bold LCC indicators (four indicators), italic LCA indicators (13 indicators), normal S-LCA indicators (26)

- Unemployment
- Fair salary

- -

- *Ionizing radiation*
- *Ozone depletion*
- *Particulate matter*
- Health expenditure
- Safety measures

- Sanitation coverage
- *Human toxicity cancer*
- *Human toxicity non-cancer*
- *Photochemical ozone creation*
- Social security expenditures

- Illiteracy, total
- Youth illiteracy, total

- Women in the sectoral labour force
- Gender wage gap

- Drinking water coverage
- Sanitation coverage

- **Levelized cost**

- Child labour, total
- Fair salary
- Social security expenditures
- Trade unionism
- Trafficking in persons

- Goods produced by forced labour
- Frequency of forced labour
- Association and bargaining rights
- **Net present value**
- **Profitability index**
- Weekly hours of work per employee
- Violations of employment laws

- **Marginal cost**

- Indigenous rights

- -

- *Abiotic resource depletion*

- *Climate change*

- *Ecotoxicity, fw.*
- *Eutrophication, fw*
- *Eutrophication, marine*

- *Acidification*
- *Eutrophication, terrestrial*

- Assoc. + barg. rights
- Trade unionism
- Violations of employment laws and regulations

- -

$$w_i = \frac{1}{n \cdot m_n / p}$$

(1)

w_i: weighting factor of indicator i, with $\sum w_i = 1$,
n: number of goals with assigned LCSA indicators, i.e. 14
m_n: number of indicators in one goal
p: number of assigned goals

To calculate the weighting factors based on the sustainability dimensions, the number of goals with assigned LCSA indicators needs to be substituted with the number of sustainability dimensions.

3   Case Study

To test the application of the SDG-guided LCSA indicator set of already existing LCA, S-LCA and LCC are adapted in a case study. The case study comprises a comparison of three locations for hydrogen production with an advanced alkaline water electrolyser. The European countries Germany, Spain and Austria offer different opportunities for industrial hydrogen production. The LCA modelling is based on Koj et al. [16], while the LCC is taken from Kuckshinrichs et al. [17]. The S-LCA [18] is conducted using the PSILCA database [11] integrated in openLCA 1.6. The functional unit for the LCSA is 1 kg of hydrogen (30 bar) produced.

4   Discussion and Results

Here the calculated indicator weights as well as the overall result using PROMETHEE are presented and compared with indicator weights derived from the approach of equal importance of the three sustainability dimensions. The values for each LCSA indicator can be found in Annex, Table 2.

4.1   SDG-Guided Indicator Weights

The derived weighting factors for the different LCSA indicators have a wide range (Table 1). They vary between 0.006 and 0.071. The results are solely based on the numbers of indicators selected for a goal, but not on any subjective assumption on the weight of indicators. Five indicators have the highest weighting factor. These are two LCC and two LCA indicators as well as indigenous rights (human rights issues faced by indigenous people).

Table 1 LCSA indicators and their weights according to SDG equal weighting

Indicator	Goal	Weight	Indicator	Goal	Weight
Child labour, total	8	0.006	Youth illiteracy, total	4	0.024
Frequency of forced labour	8	0.006	Ecotoxicity, freshwater	14	0.024
Goods produced by forced labour	8	0.006	Eutrophication, freshwater	14	0.024
Trafficking in persons	8	0.006	Eutrophication, marine	14	0.024
Net present value	8	0.006	Water depletion	6	0.024
Weekly hours of work per employee	8	0.006	Association and bargaining rights	8,16	0.015
Profitability index	8	0.006	Trade unionism	16, 8	0.015
Photochemical ozone formation	3	0.007	Violations of employment laws and regulations	8, 16	0.015
Health expenditure	3	0.007	Sanitation coverage	3, 6	0.015
Non-fatal accidents	3	0.007	Gender wage gap	5	0.036
Safety measures	3	0.007	Unemployment	1	0.036
Human toxicity, cancer	3	0.007	Women in the sectoral labour force	5	0.036
Ionizing radiation	3	0.007	Acidification, terrestrial	15	0.036
Human toxicity, non-cancer	3	0.007	Eutrophication, ter.	15	0.036
Ozone depletion	3	0.007	Fair salary	1, 8	0.021
Particulate matter	3	0.007	Indigenous rights	10	0.071
Fatal accidents	3	0.007	Levelized cost	7	0.071
Social security expenditures	3, 8	0.007	Marginal cost	9	0.071
Drinking water coverage	6	0.024	Climate change	13	0.071
Education	4	0.024	Resource depletion	12	0.071
Illiteracy, total	4	0.024			

In the approach of equal sustainability dimensions, all LCC indicators have a weight of 0.083, all LCA indicators of 0.024 and all S-LCA indicators of 0.014. With the switch from dimensions to SDGs the indicator indigenous rights shows the highest increase in the weighting factor from 0.014 to 0.071. The largest decrease is recorded for the indicator net present value from 0.083 to 0.006.

4.2   PROMETHEE Results

The PROMETHEE results for the two different weighting sets are presented in Fig. 3. High results indicate the preferable outcome. In both versions, the Spanish option is identified as the least favourable one. Both weighting sets, however, lead to different results for the most preferable option. The set based on SDGs identifies Austria as the most sustainable country for hydrogen production, while an equal weighting of the sustainability dimensions leads to the conclusion that Germany is the most preferable one.

Germany shows the best results with regard to its LCC indicators (Annex, Table 2). As these indicators lose weight (in total 0.155 instead of 0.333), Germany is not considered as the most sustainable option when SDG-guided weighting is considered. The overall weight of the LCA indicators keeps relatively constant (0.352 instead of 0.333), while the social indicators gain influence (0.420 of instead 0.333).

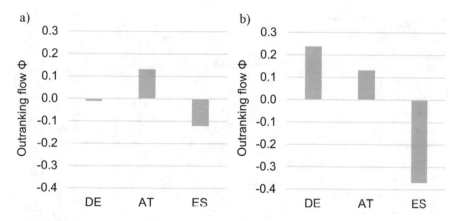

Fig. 3 PROMETHEE results of hydrogen production in three different countries: (**a**) based on SDGs, (**b**) based on sustainability dimensions (DE, Germany; AT, Austria; ES, Spain)

Table 2 LCSA indicator results, based on [5] (med. rh: medium-risk hours)

Indicator	Unit	Germany	Austria	Spain
Child labour, total	Med. rh	0.98	1.08	0.60
Frequency of forced labour	Med. rh	0.46	0.57	0.16
Goods produced by forced labour	Med. rh	0.30	0.29	0.22
Trafficking in persons	Med. rh	2.30	2.81	1.34
Weekly hours of work per employee	Med. rh	0.26	0.48	0.45
Net present value	m€$_{2015}$/kg H$_2$	−50.1	−58.1	−59.4
Profitability index	Med. rh	−6.38	−7.45	−7.74
Fatal accidents	Med. rh	0.40	0.55	0.26
Health expenditure	Med. rh	6.07	6.24	3.59
Non-fatal accidents	Med. rh	4.03	13.82	27.12
Safety measures	Med. rh	4.89	5.71	5.15
Human toxicity, non-cancer	nCTUh	37.5	14.8	27.1
Human toxicity, cancer	100 nCTUh	9.77	5.07	4.34
Ionizing radiation	100 Bq U235 eq	27.6	0.33	32
Ozone depletion	ng CFC-11 eq	6.32	4.38	5.03
Particulate matter	100 mg PM$_{2.5}$ eq	20	8.7	24.6
Photochemical ozone formation	g NMVOC	30	16.4	33

(continue)

Table 2 (continue)

Indicator	Unit	Germany	Austria	Spain
Social security expenditures	Med. rh	5.79	5.72	2.62
Drinking water coverage	Med. rh	2.60	2.90	1.65
Education	Med. rh	3.01	2.32	4.56
Illiteracy, total	Med. rh	4.45	4.43	2.21
Youth illiteracy, total	Med. rh	0.75	0.81	0.45
Ecotoxicity, freshwater	CTUe	5.59	3.31	3.71
Eutrophication, freshwater	10 mg P eq	12.8	13.3	9.32
Eutrophication, marine	m g N-eq	11.2	7.31	11.6
Water depletion	m^3 world eq.	22	22.3	43.1
Assoc. + barg. Rights	Med. rh	6.54	16.48	1.81
Trade unionism	Med. rh	25.75	18.46	43.89
Violations of employment laws and regulations	Med. rh	1.93	3.22	3.04
Sanitation coverage	Med. rh	13.89	14.17	8.15
Gender wage gap	Med. rh	5.47	31.94	7.96
Unemployment	Med. rh	0.81	0.77	37.43
Women in the sectoral labour force	Med. rh	1.85	1.93	3.93
Acidification	mMole H^+ eq.	44.5	21.6	50.3
Eutrophication, terrestrial	10 mMole N eq.	11.6	6.5	12.1
Fair salary	Med. rh	5.46	7.73	2.30
Indigenous rights	Med. rh	1.44	1.79	0.78
Levelized cost of hydrogen	$€_{2015}$/kg H_2	3.64	4.22	4.31
Marginal cost	$€_{2015}$/kg H_2	3.72	4.52	4.73
Climate change	kg CO_2 eq	29.8	29.8	29.8
Resource depletion	10 mg Sb eq	12.9	3.88	9.38

5 Conclusions

In this work, two different approaches how to cluster sustainability indicators are presented. The results show that the method considered can have a significant influence on the overall preference of options. In the case of hydrogen production in Europe, the classification based on the SDGs prefers a location in Austria, while the other classification based on the dimensions of sustainability results in a preference for a German location.

Using the same indicator set, other classifications are possible. In this paper, the dimensions of sustainability are separated by different methods. The indicators, however, can also be classified by other ways of argumentation. This could mean that human health indicators are assigned to the social dimension [e.g. 19]. In many cases, the three dimensions of sustainability are used [6]. There are other approaches available like the proposed SDGs that have different implications. For example, the SDGs do not cover indicators assessing corruption, and the stakeholder group consumers are not represented. In addition, the focus of the SDGs is less on the economic indicators and more on the social ones. In contrast, regarding the three dimensions of sustainability, some indicators might be assigned to different dimensions, e.g. resource depletion.

Another way to establish weighting factors for MCDA is not to derive them from concepts, but to ask stakeholders, e.g. residents and users or LCSA practitioners, about their preferences. It is to be expected that such an approach would probably lead to a different weighting set than the one presented. Here, the social indicator with the highest weighting factor is indigenous rights. Even though this is a very important topic, in the context of hydrogen production in three different European locations, it is probably not the most pressing social issue. Consequently, several questions arise that need to be answered in the future. An important one will be how sustainability is understood in LCSA and which principles should be at the basis?

References

1. UN General Assembly. (2015). *Resolution adopted by the General Assembly on 25 September 2015: Transforming our world: The 2030 Agenda for Sustainable Development.* United Nations.
2. https://lca-net.com/projects/show/linking-the-un-sdgs-to-life-cycle-impact-pathway-frameworks/. Accessed 31 Jan 2020.
3. Weidema, B. (2018). *Relating the UN sustainable development goals to social LCA indicators, 70th LCA discussion forum on social LCA.* Zurich.
4. Owsianiak, M., Laurent, A., Marcher, J. L., Hansen, S. L., Dong, Y., & Hauschild, M. (2019). *Indicators of sustainable development goals (SDG) as gauges of environmental sustainability, 9th international conference on life cycle management.* Poznan.
5. Wulf, C., Werker, J., Zapp, P., Schreiber, A., Schlör, H., & Kuckshinrichs, W. (2018). Sustainable development goals as a guideline for Indicator selection in life cycle sustainability assessment. *Procedia CIRP, 69,* 59–65.
6. Wulf, C., Werker, J., Ball, C., Zapp, P., & Kuckshinrichs, W. (2019). Review of sustainability assessment approaches based on life cycles. *Sustainability, 11*(20), 5717.
7. Campos-Guzmán, V., García-Cáscales, M. S., Espinosa, N., & Urbina, A. (2019). Life cycle analysis with multi-criteria decision making: A review of approaches for the sustainability evaluation of renewable energy technologies. *Renewable and Sustainable Energy Reviews, 104,* 343–366.
8. EU-JRC. (2011). *Recommendations for life cycle impact assessment in the European context - based on existing environmental impact assessment models and factors, international reference life cycle data system (ILCD) handbook.* European Commission-Joint Research Centre - Institute for Environment and Sustainability.
9. Jolliet, O., Antón, A., Boulay, A.-M., Cherubini, F., Fantke, P., Levasseur, A., McKone, T. E., Michelsen, O., Milà i Canals, L., Motoshita, M., Pfister, S., Verones, F., Vigon, B., & Frischknecht, R. (2018). Global guidance on environmental life cycle impact assessment indicators: Impacts of climate change, fine particulate matter formation, water consumption and land use. *The International Journal of Life Cycle Assessment, 23*(11), 2189–2207.
10. Andrews, E. S., Barthel, L.-P., Beck, T., Norris, C. B., Ciroth, A., Cucuzzella, C., Gensch, C.-O., Hébert, J., Lesage, P., Manhart, A., Mazeau, P., Mazijn, B., Methot, A.-L., Moberg, Å., Norris, G., Parent, J., Prakash, S., Reveret, J.-P., Spillemaeckers, S., Ugaya, C. M. L., Valdivia, S., & Weidemann, B. (2009), *Guidelines for social life cycle assessment of products: Social and socio-economic LCA guidelines complementing environmental LCA and life cycle cost-*

ing, contributing to the full assessment of goods and services within the context of sustainable development. C. B. Norris & B. Mazijn (Eds.). United Nations Environment Programme.

11. Eisfeldt, F., & Ciroth, A. (2017). *PSILCA – A product social impact life cycle assessment database version 2.* GreenDelta.

12. European Investment Bank. (2013). *The economic appraisal of investment projects at the EIB.* European Investment Bank.

13. United Nations. (2017). The Sustainable Development Goals: 17 Goals to Transform our World, Copyright © United Nations 2017. All rights reserved. Available at: http://www.un.org/sustainabledevelopment/sustainable-development-goals/. This image is distributed under the terms of the Creative Commons Attribution Non-Commercial 4.0 International licence (CC-BY-NC), a copy of which is available at http://creativecommons.org/licenses/by-nc/4.0/. United Nations.

14. Brans, J. P., Vincke, P., & Mareschal, B. (1986). How to select and how to rank projects: The Promethee method. *European Journal of Operational Research, 24*(2), 228–238.

15. Wulf, C., Werker, J., Zapp, P., Schreiber, A., & Kuckshinrichs, W. (2019). *Indikatorspezifische Indifferenzzonen in PROMETHEE für Life Cycle Sustainability Assessment der Wasserstoffproduktion, Workshop "Prospektiven multidimensionalen Bewertung von Energietechnologien".* Oldenburg.

16. Koj, J. C., Wulf, C., Schreiber, A., & Zapp, P. (2017). Site-dependent environmental impacts of industrial hydrogen production by alkaline water electrolysis. *Energies, 10*(7), 860.

17. Kuckshinrichs, W., Ketelaer, T., & Koj, J. C. (2017). Economic analysis of improved alkaline water electrolysis. *Frontiers in Energy Research, 5*, 1.

18. Werker, J., Wulf, C., & Zapp, P. (2019). Working conditions in hydrogen production: A social life cycle assessment. *Journal of Industrial Ecology, 23*(5), 1052–1061.

19. Neugebauer, S., Martínez-Blanco, J., Scheumann, R., & Finkbeiner, M. (2015). Enhancing the practical implementation of life cycle sustainability assessment – Proposal of a tiered approach. *Journal of Cleaner Production, 102*, 165–176.

Open Access This chapter is licensed under the terms of the Creative Commons Attribution 4.0 International License (http://creativecommons.org/licenses/by/4.0/), which permits use, sharing, adaptation, distribution and reproduction in any medium or format, as long as you give appropriate credit to the original author(s) and the source, provide a link to the Creative Commons license and indicate if changes were made.

The images or other third party material in this chapter are included in the chapter's Creative Commons license, unless indicated otherwise in a credit line to the material. If material is not included in the chapter's Creative Commons license and your intended use is not permitted by statutory regulation or exceeds the permitted use, you will need to obtain permission directly from the copyright holder.

Role of Stochastic Approach Applied to Life Cycle Inventory (LCI) of Rare Earth Elements (REEs) from Secondary Sources Case Studies

Dariusz Sala and Bogusław Bieda

Abstract Monte Carlo (MC) simulation using Crystal Ball® (CB) software is applied to life cycle inventory (LCI) modelling under uncertainty. Input data for all cases comes from the ENVIREE (ENVIronmentally friendly and efficient methods for extraction of Rare Earth Elements), i.e. from secondary sources eco-innovative project within the second ERA-NET ERA-MIN Joint Call Sustainable Supply of Raw Materials in Europe 2014. Case studies described the flotation tailings from the New Kankberg (Sweden) old gold mine and Covas (Portugal) old tungsten mine sent to re-processing/beneficiation for rare earth element (REE) recovery. In this study, we conduct the MC analysis using the CB software, which is associated with Microsoft® Excel spreadsheet model, used in order to assess uncertainty concerning cerium (Ce), lanthanum (La), neodymium (Nd) and tungsten (W) taken from Covas flotation tailings, as well as Ce, La and Nd taken from New Kankberg flotation tailings, respectively. For the current study, lognormal distribution has been assigned to La, Ce, Nd and W. In the case of Covas, the weights of each selected Ce, La, Nd and W are 32 ppm, 16 ppm, 15 ppm and 1900 ppm, respectively, whereas in the case of New Kankberg, the weights of each selected Ce, La and Nd are 170 ppm, 90 ppm and 70 ppm, respectively. For the presented case, lognormal distribution has been assigned to Ce, La, Nd and W. The results obtained from the CB, after 10,000 runs, are presented in the form of frequency charts and summary statistics. Thanks to uncertainty analysis, a final result is obtained in the form of value range. The results of this study based on the real data, and obtained using MC simulation, are more reliable than those obtained from the deterministic approach, and they have the advantage that no normality is presumed.

D. Sala (✉) · B. Bieda
Department of Management, AGH University of Science and Technology, Kraków, Poland
e-mail: Dsala@zarz.agh.edu.pl

© The Author(s) 2022
Z. S. Klos et al. (eds.), *Towards a Sustainable Future - Life Cycle Management*,
https://doi.org/10.1007/978-3-030-77127-0_10

1 Introduction

This paper presents the utility of uncertainty analysis based on the MC simulation applied to LCI modelling based on research data obtained from 2015 to 2017 as part of the ENVIREE EU-funded from the ERA-MIN programme within the second Joint Call aims at complete recovery process proposal of REEs (rare earth elements) from tailings and mining waste [1, 2].

The REEs are a group of 17 elements with similar chemical properties, including 15 in the lanthanide group, yttrium (Y) and scandium (Sc) due to their similar physical and chemical properties [1, 3]. The lanthanide elements traditionally have been divided into two groups: the light rare earth elements (LREEs), lanthanum (La) through europium (Eu) (Z = 57 through 63), and the heavy rare earth elements (HREEs), gadolinium (Gd) through lutetium (Lu) (Z = 64 through 71) [4]. Although Y is the lightest REE, it is usually grouped with the HREEs to which it is chemically and physically similar [4]. On the other hand, according to [5], REEs can be divided into three groups: LREEs, HREEs and scandium (Sc). LREEs comprise lanthanum (La), cerium (Ce), praseodymium (Pr), neodymium (Nd) and samarium (Sm), and the remaining are included in the HREEs. While Koltun and Tharumarajah [6] presented three groups of the REEs classification often used in extraction given in LREEs, lanthanum (La), cerium (Ce), praseodymium (Pr), neodymium (Nd) and promethium (Pm); medium rare earth elements (MREEs), samarium (Sm), europium (Eu) and gadolinium (Gd); and HREEs, terbium (Tb), dysprosium (Dy), holmium (Ho), erbium (Er), thulium (Tm), ytterbium (Yb), lutetium (Lu), scandium (Sc) and yttrium (Y) quoted in Australian Industry Commission documents [7]. By the way, definition of REEs found in the same Australian Industry Commission documents [7] is the following: "Group of 17 chemical elements – not rare at all; yttrium, for example is thought to be more abundant than lead. These elements were mislabelled because they were first found in truly rare minerals".

2 Uncertainty Analysis of LCI

The most popular approach for doing an uncertainty analysis in LCA is the MC approach [8], partly because it has been implemented in many of the major software programs for LCA, typically as the only way for carrying out uncertainty analysis (for instance, in SimaPro, GaBi and Brightway2 and in open LCA).

The MC technique is widely used and recommended for the inclusion of uncertainties for LCA. Typically, 1000 or 10,000 runs are done, but a clear argument for that number is not available, and with the growing size of LCA databases, an excessively high number of runs may be time-consuming [9, 10]. It is an important parameter in simulation modelling. [11] studied stochastic flow shop scheduling metaheuristic model for vessel transits in Panama Canal. It was found that using 200 replications is optimal, because the change in the 95% confidence interval width for makespan was negligible.

According to Good [12], the uncertainty exists when the probability of an event occurring is not 0 or 1. Not only statistic but also uncertainty is a fundamental element in simulation analysis and modelling. Definition of uncertainty given by Huijbregts [13] is the following: "Uncertainty is defined as incomplete or imprecise knowledge, which can arise from uncertainty in the data regarding the system, the choice of models used to calculate emissions and the choice of scenarios with which to define system boundaries, respectively", and uncertainty defined by Walker et al. [14] is as "any deviation from the unachievable ideal of completely deterministic knowledge of the relevant system". Uncertainty is to be found when a decision-maker cannot mention all possible outcomes and/or cannot attribute probabilities to the various outcomes [15]. According to [16], uncertainty analysis is another important issue in LCA, as average data is usually used without considering the associated variability, and the results can be misleading when comparing systems [16]. Deterministic approaches and the description of processes in the studies of ecological life cycle assessment do not properly reflect the reality [17]. The analysis of uncertainty, a pervasive topic in LCA studies [18, 19], has been a subject for more than 10 years. Many LCA software tools (e.g. SimaPro, GaBi) facilitate uncertainty propagation by means of sampling methods, and most often used MC simulation [16, 20–22]. Detailed description of the combination of sources of uncertainty (parameter, model and scenario uncertainties) and combination of source of uncertainty and methods to address them (deterministic, probabilistic and simple methods) are discussed in [23].

MC simulation has received considerable attention in the literature, especially when MC simulations are used for making decisions that will have a large social and economic impact [24]. As a result, it was the most commonly recommended tool (e.g [25, 26]). Stochastic nature of the MC simulation is based on random numbers, and simulation models are generally easier to understand than many analytical approaches [18]. According to La Grega et al. [27], MC simulation can be considered the most effective quantification method for uncertainties and variability among the environmental system analysis tools available.

3   LCI Data Quality and Collection

Based on the different physical and chemical separations carried out on New Kankberg and Covas tailings [28], the following process treatment scheme is shown in Figs. 1 and 2, respectively.

The possibilities of extraction of Ne, Ce and La using magnetic separation can be reached, thanks to the paramagnetic property of monazite. Inventory data used in the study has been obtained from the following sources: the primary data used in this study is based on the elements determined from the chemical analyses done by instrumental neutron activation analyses site-specific measured or calculated data,

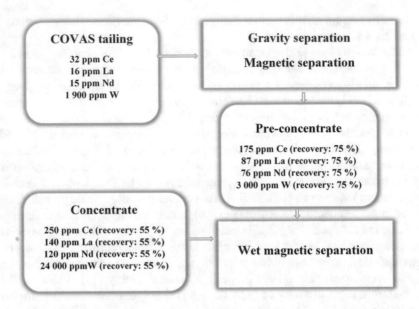

Fig. 1 Proposed process scheme for the beneficiation of Covas tailings. (Adopted from [28])

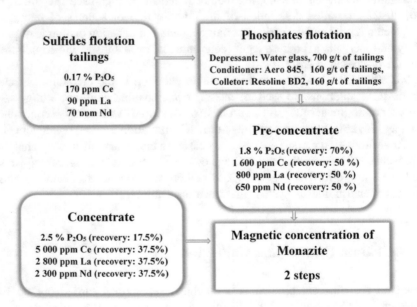

Fig. 2 Proposed process scheme for the beneficiation of REE in the flotation tailings from New Kankberg mine. (Adopted from [28])

and on values found in literature. In the current study, we discuss and model our LCI adopting the proposed process for the beneficiation of REE in the flotation tailings from New Kankberg mine in Sweden and Covas tailings [29].

4 Simulation Model: Model Assumptions

Simulation models are generally easier, when it comes to their interpretation and understanding, than a number of analytical solutions. Moreover, simulation models provide an interesting opportunity to give more reliable and comprehensive data [30]. For input parameters analysed in this study (La, Ce, Ne and W), uncertainty was included in the MC analysis by assigning distributions.

For uncertainty analysis in the LCI study, the lognormal probability distributions have been assigned to each analysed REE. Lognormal distribution is stable and no negative values are possible [21]. In this context, it should be pointed out that the lognormal probability distribution with the GSD equal to 1.13 was applied to rare earth oxides in the ecoinvent background process "Rare earth oxide production from bastnaesite" taken from the "Life Cycle Inventories of Chemicals Data v2.0 Ecoinvent report No. 8" [31].

The decision to choose lognormal distribution is based on the works of [20, 21, 32] and the bibliographies included in the above-mentioned publication because the quality of data was not sufficient to estimate best-fitting distributions.

Several examples of performance of MC simulation by using CB software can be found in [33] as well as in [20, 21, 34]. The MC simulation results for La, Ce, Ne and W are shown in graphical forms (histograms) and descriptive statistics (percentiles summary and statistics summary).

It is important that a sufficient number of replications (runs) should be used in a simulation [35], because the quality of the simulation results depends on the number of replications. In general, the higher the number of replications, the more accurate will be the characterization of the output distribution and estimates of its parameters, such as the mean [34].

5 Results and Discussion

Random values from the probability distribution of each parameter were selected in each run and a forecast distribution for each selected REE. CB's distribution fitting function can analyse a data set and determine not only the best fit but also the quality of the fit [34]. During a single trial, CB randomly selects a value from the defined possibilities (the range and shape of the distribution) for each uncertain variable and then recalculates the spreadsheet [36].

5.1 Covas (Portugal) Old Tungsten Mine Case Study

After activating the simulation with the randomization cycle, set previously to 10,000 trials, the results obtained by MC simulation after 10,000 trials, for the Ce, La and Ne, have been presented in the form of frequency charts (histograms). They are shown in Figs. 3, 4, 5 and 6, respectively; statistics, as well as percentiles, reports are presented in Tables 1 and 2, respectively. The mean values of Ce, La, Ne and W forecast values amounted to the GSD with a 95% confidence interval around the mean values were situated between:

- Ce [26.17 and 38.61] ppm (see Fig. 3)
- La [13.13 and 19.46] ppm (see Fig. 4)
- Nd [12.24 and 18.06] ppm (see Fig. 5)
- W [1556.96 and 2302.73] ppm (see Fig. 6)

The histograms of the outcome variables include all values within 2.6 standard deviations from the mean, which represents approximately 99% of the data, and the number of data points inside 2.6 standard deviations of the mean is shown in the upper right corner of the frequency charts, as presented in Figs. 3, 4, 5 and 6 (see [20, 34] for more details). It is worth noting that if the number of runs increases, the mean standard error decreases [34]. Moreover, the mean standard error can be used to construct confidence intervals as described in Evans and Olson [34].

The confidence interval range expressing 95% presented in the frequency chart (see Figs. 3, 4, 5 and 6) is highlighted with a darker colour marker. In other words, this means that 95% of the results are lying inside this range. Moreover, by setting the certainty values (e.g. 95%), the confidence intervals (minimum and maximum bounds) are set automatically by the grabbers, and the corresponding numerical values are entered in the edit fields at the bottom part of the dialog boxes of the Forecast tab (e.g [20, 34].).

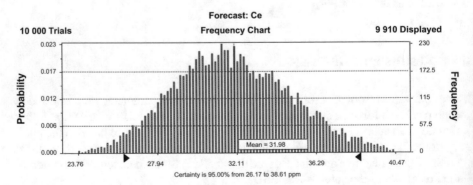

Fig. 3 CB forecast chart: Ce after 10,000 trials (95% confidence interval). Certainty is 95.00% from 26.17 to 38.61 ppm. (Source: own work)

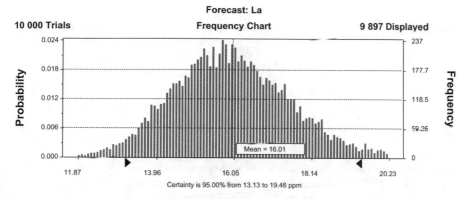

Fig. 4 CB forecast chart: La after 10,000 trials (95% confidence interval). Certainty is 95.00% from 13.13 to 19.46 ppm. (Source: own work)

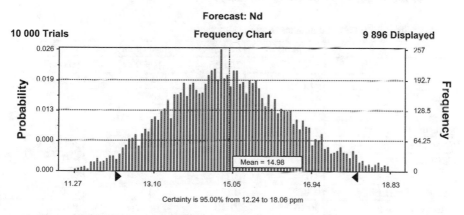

Fig. 5 CB forecast chart: Nd after 10,000 trials (95% confidence interval). Certainty is 95.00% from 12.24 to 18.06 ppm. (Source: own work)

5.2 New Kankberg (Sweden) Old Gold Mine Case Study

The results obtained by MC simulation, after 10,000 runs, for Ce, Ne and La, are shown in Figs. 7, 8 and 9, respectively, as well as in statistics and percentiles reports presented in Tables 3 and 4, respectively. The mean values of Ce, Nd and La with a 95% confidence interval around the mean values were situated between:

- Ce [138.93 and 207.00] ppm (see Fig. 7)
- Nd [57.29 and 84.67] ppm (see Fig. 9)
- La [73.97 and 108.33] ppm (see Fig. 8)

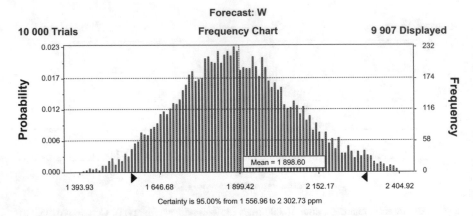

Fig. 6 CB forecast chart: W after 10,000 trials (95% confidence interval). Certainty is 95.00% from 1556.96 to 2302.73 ppm. (Source: own work)

Table 1 Statistics report of outcomes from the simulation

Statistic	Ce (ppm)	La (ppm)	Ne (ppm)	W (ppm)
Trials	10,000	10,000	10,000	10,000
Mean	31.98	16.01	14.98	1898.60
Median	31.80	15.93	14.91	1887.51
Mode	–	–	–	–
Standard deviation	3.19	1.60	1.49	191.02
Variance	10.19	2.56	2.22	36487.08
Skewness	0.27	0.33	0.25	0.31
Kurtosis	3.05	3.23	3.02	3.13
Coeff. Of variability	0.10	0.10	0.10	0.10
Range maximum	36.46	21.10	19.47	1904.42
Range minimum	20.36	10.26	9.66	1284.84
Range width	47.24	23.51	20.81	2692.45
Mean std. error	0.03	0.02	0.01	1.91

Source: own work

6 Conclusions

This study provides new insight into the practical implementation of MC method, based on the stochastic approach, and applied to the uncertainty of the LCI data collection process. To our knowledge, there is a lack of publications and research presentation of stochastic modelling of the data used for the LCI, for beneficiation of REEs, in the flotation tailings processes. Probabilistic techniques using MC simulations must consider the strategy based on the specification of the optimal distribution. The MC simulation in this study provides justification for the lognormal distributions assumed for the analysed parameters. Thanks to uncertainty analysis, a final result is obtained in the form of value range. As a result, the results of this

Table 2 Percentiles report of outcomes from the simulation

Percentile	Ce (ppm)	La (ppm)	Ne (ppm)	W (ppm)
0%	20.36	10.26	9.66	1284.84
10%	28.04	14.03	13.13	1659.92
20%	29.26	14.65	13.71	1736.95
30%	30.17	15.13	14.14	1793.51
40%	31.01	15.53	14.56	1841.49
50%	31.80	15.93	14.91	1887.51
60%	32.62	16.33	15.29	1938.46
70%	33.61	16.78	15.71	1900.61
80%	32.94	16.95	15.90	1989.04
90%	36.19	18.11	16.92	2147.66
100%	47.24	23.51	20.81	2692.45

Source: own work

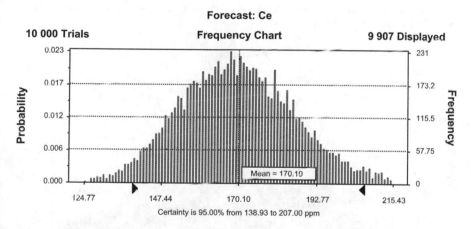

Fig. 7 CB forecast chart: Ce after 10,000 trials (95% confidence interval). Certainty is 95.00% from 138.93 to 207.00 ppm. (Source: own work)

study, based on the real data and obtained using MC simulation, are more reliable than those based on the deterministic approach. An additional advantage is associated with the fact that no normality is presumed.

Finally, it is concluded that uncertainty analysis offers a well-defined procedure for LCI studies, early phase of LCA, and provides the basis for defining the data needs for full LCA of the beneficiation of REE process. It must be pointed out that MC simulation needs to know the probability distribution for the purpose of an uncertainty analysis in contrast to bootstrap sampling, which creates an uncertainty analysis without knowing the probability distribution of the analysed data.

Stochastic approach used to LCI supports decision-makers in the interpretation of final LCA results and leads to better understanding of many analytical approaches. The results of this study will encourage other researchers to consider this approach in their projects. Results can improve current procedures, and they can help the

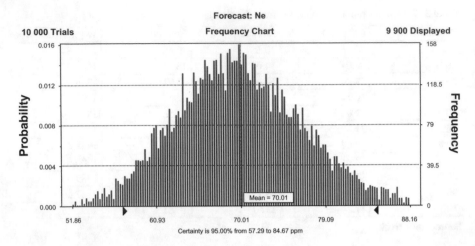

Fig. 8 CB forecast chart: Ne after 10,000 trials (95% confidence interval). Certainty is 95.00% from 57.29 to 84.67 ppm. (Source: own work)

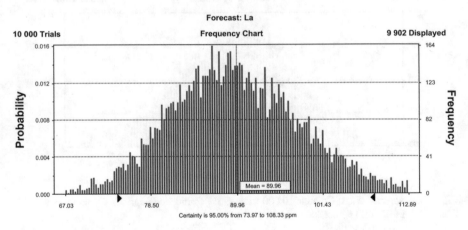

Fig. 9 CB forecast chart: La after 10,000 trials (95% confidence interval). Certainty is 95.00% from 73.97 to 108.33 ppm. (Source: own work)

LCA practitioners and decision-makers in the REEs beneficiation processes modelling and management. They can also contribute to better understanding of many analytical procedures and bring closer to industrial application – industrially relevant focus – and may also stimulate innovation in the stochastic studies.

Summarizing, consideration of uncertainty in LCA will make the LCA field more robust and credible in supporting the practitioner decisions, as discussed in the work of Igos et al. [10].

Table 3 Statistics report of outcomes from the simulation

Statistic	Ce (ppm)	La (ppm)	Ne (ppm)
Trials	10,000	10,000	10,000
Mean	170.19	89.96	70.01
Median	169.42	89.44	69.67
Mode	....	....	....
Standard deviation	17.23	8.82	6.98
Variance	296.77	77.77	48.75
Skewness	0.30	0.28	0.29
Kurtosis	3.18	3.10	3.16
Coeff. of variability	0.10	0.10	0.10
Range maximum	245.86	131.45	103.56
Range minimum	113.19	62.87	43.33
Range width	132.67	68.58	60.23
Mean std. error	0.17	0.09	0.07

Source: own work

Table 4 Percentiles report of outcomes from the simulation

Percentile	Ce (ppm)	La (ppm)	Ne (ppm)
0%	113.19	62.87	43.33
10%	148.78	79.10	61.20
20%	155.45	82.45	64.07
30%	160.49	85.00	66.17
40%	165.08	87.28	67.95
50%	169.42	89.44	69.67
60%	173.82	91.70	71.40
70%	178.62	94.38	73.44
80%	184.30	97.25	75.81
90%	192.43	101.51	79.09
100%	245.86	131.45	103.56

Source: own work

Acknowledgements The authors are grateful for the input data provided, as part of the environmentally friendly and efficient methods, for extraction of rare earth elements from secondary sources (ENVIREE) project funded by NCBR within the second ERA-NET ERA-MIN Joint Call Sustainable Supply of Raw Materials in Europe 2014.

Funding This work was supported by the Management Department of the AGH University of Science and Technology, Kraków, Poland.

Compliance with ethical standards. *Conflict of interest*: The authors declare that they have no conflict of interest. Research is not involving human participants and/or animals.

References

1. Grzesik, K., Bieda, B., Kozakiewicz, R., & Kossakowska, K. (2017). Goal and scope and its evolution for life cycle assessment of rare earth elements recovery from secondary sources. *SGEM 2017 Geoconference: Energy and Clean Technologies Albena.*, Nuclear technologies recycling air pollution and climate change, *17*(41), 107–114.
2. ENVIREE. (2015). http://www.enviree.eu/home. Accessed 22 Feb 2020.
3. Navarro, J., & Zhao, F. Life-cycle assessment of the production of rare-earth elements for energy applications: A review. *Frontiers in Energy Research.* https://doi.org/10.3389/fenrg.2014.00045. Accessed 22 Feb 2020.
4. Castor, S. B., & Hedric, J. B. (2006). *Rare earth elements, industrial minerals and rocks* (7th ed.). Society for Mining, Metallurgy and Exploration.
5. Gutiérrez-Gutiérrez, S. C., Coulon, F., Jiang, Y., & Wagland, S. (2015). Rare earth elements and critical metal content of extracted landfilled material and potential recovery opportunities. *Waste Management, 42*, 128–136.
6. Koltun, P., & Tharumarajah, A. Life cycle impact of rare earth elements. *ISRN Metallurgy*, Article ID 907536. https://doi.org/10.1155/2014/907536. Accessed 21 Feb 2020.
7. Australian Industry Commission, New and Advanced Materials, Australian Government Publishing Service, Melbourne, Australia. https://www.pc.gov.au/inquiries/completed/new-advanced-materials/42newmat.pdf. Accessed 21 Feb 2020.
8. Lloyd, S. M., & Ries, R. (2007). Characterizing, propagating and analyzing uncertainty in life-cycle assessment. A survey of quantitative approaches. *Journal of Industrial Ecology, 11*, 161–179.
9. Heijungs, R. (2020). On the number of Monte Carlo runs in comparative probabilistic LCA. *The International Journal of Life Cycle Assessment, 25*, 394–402.
10. Igos, E., Benetto, E., Meyer, R., Baustert, P., & Othoniel, B. (2019). How to treat uncertainties in life cycle assessment studies? *The International Journal of Life Cycle Assessment, 24*(4), 794–807.
11. Jackman, J., Guerra de Castillo, Z., & Olafsson, S. (2011). Stochastic flow shop scheduling model for the Panama Canal. *Journal of the Operational Research Society, 62*, 69–80.
12. Good, I. J. (1995). Reliability always depends on probability of course. *Journal of Statistical Computation and Simulation, 52*, 192–193.
13. Huijbregts, M. A. J. (1998). Application of uncertainty and variability in LCA. Part I: A general framework for the analysis of uncertainty and variability in life cycle assessment. *International Journal of Life Cycle Assessment, 3*(5), 273–280.
14. Walker, W. E., Harremoës, P., Rotmans, J., van der Sluijs, J. P., van Asselt, M. B. A., Janssen, P., & Krayer von Krauss, M. P. (2003). Defining uncertainty: A conceptual basis for uncertainty management in model-based decision support. *Integrated Assessment, 4*(1), 5–17.
15. Thomas, C. T., & Maurice, S. C. *Decisions under risk and uncertainty*. Managerial Economics. http://highered.mheducation.com/sites/0070601607/student_view0/chapter15/index.html. Accessed 22 Mar 2018.
16. Escobar, N., Ribal, J., Clemente, G., & Sanjuán, N. (2014). Consequential LCA of two alternative systems for biodiesel consumption in Spain, concerning uncertainty. *Journal of Cleaner Production, 79*, 61–73.
17. Canarache, A., Simota, C., et al. (2002). In M. Pagliai & R. Jones (Eds.), *Sustainable land management-environmental protection, a soil physical approach* (Advances in geoecology 35) (pp. 495–506). Catena Verlag GmbH.
18. Heijungs, R., & Lenzen, M. (2014). Error propagation methods for LCA – A comparison. *International Journal of Life Cycle Assessment, 19*, 1445–1461.

19. Heijungs, R. (2020). On the number of Monte Carlo runs in comparative probabilistic LCA. *The International Journal of Life Cycle Assessment, 25*, 394–402.
20. Bieda, B. (2012). *Stochastic analysis in production process and ecology under uncertainty.* Springer-Verlag.
21. Sonnemann, G., Castells, F., & Schumacher, M. (2004). *Integrated life-cycle and risk assessment for industrial processes.* Lewis Publishers.
22. Escobar, N., Ribal, J., Clemente, G., Rodrigo, A., Pascual, A., & Sanjuán, N. (2015). Uncertainty analysis in the financial assessment of an integrated management system for restaurant and catering waste in Spain. *The International Journal of Life Cycle Assessment, 20*, 491–1510.
23. Scope, C., Ilg, P., Muench, S., & Guenther, E. J. (2016). Uncertainty in life cycle costing for long-range infrastructure. Part II: guidance and suitability of applied methods to address uncertinty. *The International Journal of Life Cycle Assessment, 21*, 1170–1184.
24. Saltelli, A., Tarantola, S., Campolongo, F., & Ratto, M. (2004). *Sensitivity analysis in practice. A guide to assessing scientific models.* Wiley.
25. Guo, M., & Murphy, R. J. (2012). LCA data quality: Sensitivity and uncertainty analysis. *Science of the Total Environment, 435–436*, 230–243.
26. Skalna, I., Rębiasz, B., Gaweł, B., Basiura, B., Duda, J., Opiła, J., & Pełech-Pilichowski, T. (2015). *Advances in fuzzy decision making, studies in fuzziness and soft computing 333.* Springer Verlag.
27. LaGrega, M. D., Buckingham, P. L., & Evans, J. C. (1994). *Hazardous Waste Management.* Mc Graw-Hill.
28. Mcnard, Y., & Magnaldo, A. *ENVIREE DELIVERABLE D2.1: Report on the most suitable combined pre-treatment, leaching and purification processes.* http://www.cnviree.eu/fileadmin/user_upload/ENVIREE_D2.1.pdf. Accessed 21 Feb 2020.
29. Marques Dias, M. I, Borcia, C. G., & Menard, Y. *ENVIREE – D1,2 and D1.3 reports on properties of secondary REE sources.* http://www.enviree.eu/fileadmin/user_upload/ENVIREE_D1.2_and_D1.3.pdf. Accessed 21 Feb 2020.
30. Rönnlund, I., Reuter, M., Horn, S., Aho, J., Aho, M., Päällysaho, M., Ylimäki, L., & Pursula, T. (2016). Eco-efficiency indicator framework implemented in the metallurgical industry: Part 1- a comprehensive view and benchmark. *The International Journal of Life Cycle Assessment, 21*, 1473–1500.
31. Althaus, H-J., Hischier, R., Osses, M., Primas, A., Hellweg, S., Jungbluth, N., & Chudacoff, M. *Life cycle inventories of chemicals data v2.0 Ecoinvent report no. 8.* Dübendorf, https://db.ecoinvent.org/reports/08_Chemicals.pdf. Accessed 28 Feb 2020.
32. Muller, S., Lesage, P., Ciroth, A., Mutel, C., Weidema, B. P., & Samson, R. (2016). The application of the pedigree approach to the distributions foreseen in ecoinvent v3. *International Journal of Life Cycle Assessment, 21*, 1327–1337.
33. Gonzalez, A. G., Herrador, M., & Asuero, A. G. (2005). Uncertainty evaluation from Monte-Carlo simulations by using Crystal-Ball software. *Accreditation and Quality Assurance, 10*, 149–154.
34. Evans, J. R., & Olson, D. L. (1998). *Introduction to simulation and risk analysis.* Prentice Hall. Inc. A Simon & Schuster Company.
35. Warren-Hicks, W. J., & Moore, D. R. (1998). Uncertainty analysis in ecological risk assessment. In *Proceeding from the Pellston workshop on uncertainty analysis in ecological risk assessment*, 23–28 August 1995. Society of Environmental Toxicology and Chemistry/SETAC, Pellston, Michigan, Pensacola, FL.
36. Risk Analysis Overview. https://www.crystalballservices.com/Portals/0/eng/risk-analysis-overview.pdf?ver=2013-11-14-135039-623. Accessed 21 Feb 2020.

Open Access This chapter is licensed under the terms of the Creative Commons Attribution 4.0 International License (http://creativecommons.org/licenses/by/4.0/), which permits use, sharing, adaptation, distribution and reproduction in any medium or format, as long as you give appropriate credit to the original author(s) and the source, provide a link to the Creative Commons license and indicate if changes were made.

The images or other third party material in this chapter are included in the chapter's Creative Commons license, unless indicated otherwise in a credit line to the material. If material is not included in the chapter's Creative Commons license and your intended use is not permitted by statutory regulation or exceeds the permitted use, you will need to obtain permission directly from the copyright holder.

Extending LCA Methodology for Assessing Liquid Biofuels by Phosphate Resource Depletion and Attributional Land Use/Land Use Change

Heiko Keller, Horst Fehrenbach, Nils Rettenmaier, and Marie Hemmen

Abstract Many pathways towards reaching defossilization goals build on a substantially increased production of bio-based products and energy carriers including liquid biofuels. This is, amongst others, limited by land and phosphorous availability. However, it is challenging to adequately capture these limitations in LCA using state-of-the-art LCI and LCIA methods. We propose two new methods to overcome these challenges: (1) attributional land use and land use change (aLULUC) evenly attributes LU-/LUC-related burdens (emissions) occurring in a country to each hectare of cropland used in that country and (2) phosphate rock demand as a stand-alone resource indicator for a finite resource that cannot be replaced. Approach, calculations and used factors are described for both methods, and exemplary results for biofuels are presented. We conclude that both methods can yield additional insight and can support finding solutions for current challenges in agriculture.

1 Introduction

As for most bio-based products, replacing fossil fuels by biofuels mostly creates environmental advantages and disadvantages at the same time. Advantages typically relate to climate change mitigation and savings of fossil energy resources, and disadvantages of various kinds are usually caused by the required biomass production. This well-known pattern is reflected in standard LCA results in the field.

Public and scientific discussions more and more focus on environmental burdens and limitations of agriculture that are becoming important bottlenecks of agriculture on a global scale. These aspects include land use/biodiversity, water and increasingly also limited phosphate resources. These could also become limiting for currently discussed pathways for defossilization of the society, which often builds on using more bio-based products in general and biofuels of various kinds in particular.

H. Keller (✉) · H. Fehrenbach · N. Rettenmaier · M. Hemmen
ifeu – Institute for Energy and Environmental Research Heidelberg, Heidelberg, Germany
e-mail: Heiko.Keller@ifeu.de

© The Author(s) 2022
Z. S. Klos et al. (eds.), *Towards a Sustainable Future - Life Cycle Management*,
https://doi.org/10.1007/978-3-030-77127-0_11

Results of state-of-the-art LCAs however often do not effectively support finding new solutions in these areas for various reasons. This paper focusses on the aspects land use/land use change (LUC) and phosphate resources. In the following chapters, limitations of current state-of-the-art LCA methods are discussed, and two new methods are proposed as solutions: (1) attributional land use and land use (aLU-LUC) change as new alternative to dLUC/iLUC and (2) phosphate rock demand as new stand-alone resource indicator.

2 Attributional Land Use and Land Use Change

2.1 Background

Land use change (LUC) describes the relative change in the use or management of an area compared to a previous use of the same area and the associated emissions (or emission avoidance). Which methodology is suitable for the quantification LUC-related burdens depends on the goal and scope of a study. This can include the overall greenhouse gas balance of a country, the traceable direct consequences of a specific product in its supply chain (dLUC, direct land use change) or the indirect consequences of a change in the market, e.g. triggered by the support of a specific product such as biofuels (iLUC, indirect land use change).

In theory, dLUC could accurately determine the actual LUC emissions from a product such as rapeseed diesel. However, this is not applicable in practice for several reasons: Firstly, existing data is not available and subject to data protection. Secondly, more biomass not associated with land use change is available than interested customers or regulated markets are demanding. Thus, dLUC is not useful to mitigate or stop continuing land use change.

iLUC factors are calculated by combining land use models with an economic equilibrium or partial system and are intended to estimate the overall impact of a targeted or shock-like increase in production on global land use. Fehrenbach [18], amongst others, analysed and described the wide range of results depending on the choice of model. The iLUC approach is therefore only of limited use for developing solutions based on life cycle assessments due to the disagreement amongst experts about the suitability and reliability of the various iLUC models. Moreover, iLUC always describes results of changes or measures, which is incompatible with attributional LCAs describing the status quo. Finkbeiner [19] also discussed these aspects in detail.

We propose a life cycle inventory approach termed attributional land use/land use change (aLULUC) to attribute existing and documented burdens caused by land use change and continuous burdens/emissions from using converted land to products [1]. Here we focus on climate impacts although further impact categories such as biodiversity [2, 3] can also be assessed using the life cycle inventory method aLULUC.

2.2   Approach

A decisive premise for aLULUC is that land use changes to arable land take place in reality. These land use changes are usually recorded and associated emissions are backed up with data. This includes one-time emissions from actual LUC and continuous emissions mainly from organic soils caused by LUC but occurring for many decades of land use (LU) that can only be stopped if land use is given up and appropriate protection measures are taken.

The aLULUC concept is independent of models of future land use change as it is the case for iLUC. In the same systematic way as real emissions are attributed to the processes of a life cycle, real LUC processes can be attributed to the associated processes, as it is also done applying the dLUC concept. However, even if actual land use changes can be clearly assigned to certain agricultural products, all agricultural products of a production area compete for limited availabilities on the local market for cropland. The reaction of the markets on, e.g. the EU biofuels policy, has shown that crops on and products from recently logged land (or from "LUC-free" land) can be flexibly allocated to customers according to their preferences. For that reason, a land-market-based attribution of aLULUC to products produced on that land following the aLULUC concept is a more consistent representation of the underlying processes than a direct attribution following the dLUC concept. For the majority of agricultural products, country borders are the most appropriate geographical reference areas for the aLULUC concept. Firstly, there are no internal trade barriers within national markets. More importantly, however, decisions and policies regarding the conservation of areas such as rainforests, wetlands and grasslands are made or influenced at the country level. A more specific attribution of LUC to individual crops within these markets would require economic assumptions and models that seek to establish causalities. These do not necessarily reflect the complex socio-economic and political processes that can cause, promote or prevent LUC.

Following the proposed aLULUC approach, the real land use changes that have been caused by agriculture (of a defined region) are allocated to all agricultural products in proportion to the land requirements. It is therefore an allocation according to the attribute land demand. aLULUC can be calculated for arable land as well as for other types of land such as grassland. The country- and year-specific aLULUC factor for arable land is determined as follows: All carbon stock changes in biomass and soils caused by net conversion from other types of land use to arable land in a country in a certain year are summed up and divided by the area of arable land used in that year. One-time changes in biomass and soil carbon stocks (LUC) are attributed to the year in which the LUC occurs although actual CO_2 emissions may be partially delayed by a few years. Continuous emissions of CO_2, CH_4 and N_2O from the cultivation of organic soils (LU) are counted in the year in which they occur. Averages of aLULUC factors over the last ten available years result in stable values that do not disregard medium- to long-term developments. Detailed calculation procedures and data sources are discussed in [1]. Current emission factors for the climate impact of land use and land use change according to the aLULUC concept for selected countries can be found in Table 1.

2.3 Application Example: GHG Emissions Including aLULUC of European Rapeseed Biodiesel

Biodiesel can achieve certain climate change mitigation if it replaces conventional diesel. This is however only the case if land use does not cause high additional greenhouse gas emissions. Usually, this problem is discussed for palm oil biodiesel and deforestation. However, depending on the used land and the methodological approach used to attribute emissions from LU and LUC to the fuel, also European rapeseed biodiesel can cause in total more greenhouse gas emissions than it saves (Fig. 1). Greenhouse gas emissions from LU and LUC in Europe mainly stem from conversion of grassland and from cultivation on organic soils, i.e. drained wetlands/peatland.

The cultivation of rapeseed on former grassland can lead to overall additional contributions to climate change following the dLUC approach if common time horizons of up to about 25 years are used. If organic soils/peatlands are used, an analogous direct attribution of LU to the product (termed dLU in the figure) could even lead to very high additional greenhouse gas emissions. Where such emissions have

Table 1 Exemplary country-specific aLULUC emission factors for annual crops cultivated on arable land selected from [1]

Country	Total aLULUC t CO_2eq/(ha year)	aLU t CO_2eq/(ha year)	aLUC t CO_2/(ha year)
France	0.90	0.41	0.50
Germany	1.44	1.22	0.21
Italy	0.26	0.14	0.12
Netherlands	4.50	4.08	0.41
Poland	1.60	1.59	0.01
Romania	0.17	0.14	0.03
Spain	0.04	0.04	0.01
United Kingdom	0.55	0.55	0.00
EU 28	1.05	0.85	0.20
Argentina	3.36	0.03	3.33
Colombia	52.3	0.00	52.3
Brazil	9.32	0.00	9.32
Malaysia	55.4	42.9	12.5
Indonesia	30.4	13.7	16.7
India	0.06	0.06	0.00
Russia	0.80	0.30	0.50
USA	0.52	0.52	0.00

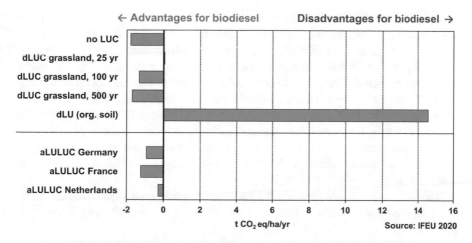

Fig. 1 Life cycle greenhouse gas emissions of European rapeseed biodiesel compared to conventional diesel. All results are based on the same life cycle comparison with differences only in the used land and the methodological approach to land use (LU) and land use change (LUC). The time horizons, over which one-time emissions are distributed, are specified where applicable

to be attributed to a fuel according to the European renewable energy directive [4], no farmer would of course cultivate crops for biofuels. Nevertheless, biofuel crops occupy land and increase the pressure to use former grassland and peatland for cultivation of crops in general. Following the aLULUC approach, LU- and LUC-related emissions are evenly distributed over all cropland of the respective country. This leads to somewhat reduced climate change mitigation for French rapeseed biodiesel. Especially emissions from cultivated organic soils in Germany and even more so in the Netherlands lead to a substantial reduction of greenhouse gas emission savings.

This application example shows that LU and LUC can make significant contributions to carbon footprints also in European countries. The aLULUC approach helps that these emissions are not neglected because direct attribution of these emissions to products does not take place in practice – neither in Europe nor overseas. Hardly anybody would, for example, consider that, e.g. his/her meat could stem from animals raised on corn grown on drained Northern European peatland.

3 Phosphate Rock Demand

3.1 Background

Phosphate rock is the basic raw material for the production of phosphoric acid, which is essential for the production of phosphate products such as fertilizers, animal feed, food and other industrial products. Ninety per cent of the global supply of phosphate is used as fertilizer in agriculture [5]. Eighty-five per cent of phosphate ore is extracted from marine sedimentary deposits and 15% from magmatic deposits, with phosphate ore chemically including iron and aluminium salts as hydrate complexes with very different phosphorous and phosphate contents. Deposits based on guano deposits are largely exhausted [6]. The main producing countries are currently China (52%), the USA (10%) and Morocco (12%) [7]. Marketable rock phosphate contains between 27% and 40% phosphate ([8] cited in [6]). Besides, recycled phosphate can be recovered from sewage sludge by several processes [9, 10].

As a mineral raw material, phosphate is a non-renewable resource. Depending on the source, the static lifetime of global phosphate reserves is only several decades to a few centuries [11–14] (see Fig. 2).

This shortage is further worsened by a growing world population and simultaneously changing consumption patterns [15], resulting in an increasing demand for phosphate.

Due to this growing importance, the impossibility of substitution by other raw materials in central applications and simultaneous limitation, we recommend integrating the resource "phosphate" in life cycle assessments using a separate indicator. We suggest using the indicator phosphate rock demand as proposed in [16] and presented below.

3.2   Approach

Various indicators can be used in LCA to address resource use. One indicator is the cumulative raw material demand (CRD), which is defined as the sum of all raw materials entering a system – except water and air – expressed in mass units. Other indicators also include weighting of the individual raw materials by, e.g. scarcity. These established indicators have in common that the mineral resource consumption of phosphate is not reported separately. This is not sufficient especially for LCAs with a strong focus on agriculture such as LCAs on biofuels because phosphate/phosphorus is a raw material that cannot be replaced by any other element in

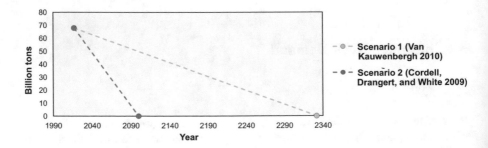

Fig. 2 Range of scenarios on the static lifetime of global phosphate reserves

its main application as a fertilizer. Therefore, the consumption of non-renewable phosphate rock needs to be addressed independently from other raw materials.

We propose a new indicator phosphate rock demand (informally also "phosphate rock footprint") following the concept of the CRD but only including phosphate rock. Phosphate rock demand is determined by the initial rock mass. The recommended unit for the life cycle inventory is "phosphate rock standard" [16]. The definition of a standard is necessary because phosphate rock can have significantly different phosphate contents. Based on [17], an average content of 25% of P_2O_5 is set for phosphate rock standard. P_2O_5 is the reference substance/unit commonly used in agriculture. This corresponds to 32% raw phosphate. This means that 1 kg of mineral P_2O_5 fertilizer corresponds to 4.0 kg of phosphate rock (std.) or 3.125 kg of raw phosphate (std.). This specification explicitly refers to mineral fertilizers. For organic fertilizers, a specific procedure must be derived depending on the goal and scope of the study. If consequential modelling is applied, for example, additional phosphate sources are taken into account, which can replace mineral phosphate without restrictions and which are available in limited quantities during the reference period of the study.

Results can be normalized to inhabitant equivalents by dividing them by the average annual resource consumption per inhabitant. The following normalization factors were derived for this purpose ([16] for details):

- For the reference area Germany: 16.1 kg phosphate rock (std.)/(inhabitant • year).
- For the reference area Europe: 23.1 kg phosphate rock (std.)/(inhabitant • year).

These factors refer to the 5-year average and thus remove short-term fluctuations in the statistics. Normalization factors for other regions can be derived accordingly.

3.3   Application Example: Phosphate Rock Demand of Different Biofuels

With the approach described in Chap. 3.2 outlining the definition and calculation of the indicator "phosphate rock demand", the resource phosphate can be integrated into life cycle assessments. In the following, the application of this approach is explained using an illustrative example. Several bio-based fuels were analysed: bioethanol, biomethane, biodiesel, fuel from vegetable oil and Fischer-Tropsch diesel. Figure 3 shows the ranges between minimum and maximum phosphate rock demand per biofuel.

The phosphate rock demand of different biofuels differs significantly. First-generation biofuels tend to perform much better than second-generation bioethanols with respect to phosphate rock depletion. Results also depend heavily on the biofuels' production schemes, co-product uses and their local conditions. This striking difference between first- and second-generation bioethanols mainly results from phosphate inputs into fermentation processes without subsequent productive

Disadvantages for biofuel →

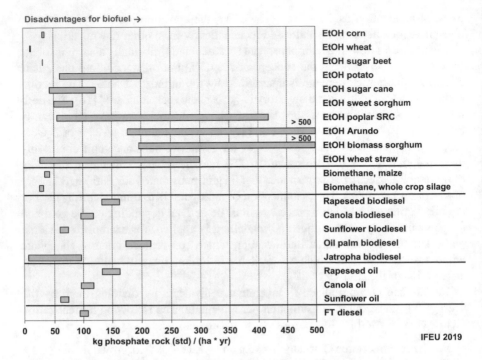

Fig. 3 Phosphate rock demand of different biofuels compared to the respective conventional fossil fuel. The ranges encompass conservatively and optimistically estimated phosphate rock demands for each fuel. EtOH stands for bioethanol, SRC for short rotation coppice and FT for Fischer-Tropsch

recovery. Based on the large range of results, it seems plausible that this aspect has not been optimized or not even been recognized as potential problem in the current state of process development and maturation. This underlines the importance of the indicator phosphate rock demand to support finding solutions to the problem of declining non-renewable phosphate resources.

4 Conclusions

In this paper, we presented two LCA extensions that intend to better address limitations of current agriculture in decision-making processes.

The life cycle inventory method attributional land use and land use change (aLU-LUC) evenly attributes impacts of deforestation, grassland conversion (both LUC) and organic soil use (LU) actually taking place in a country to each hectare of cropland used in that country. This has several advantages over commonly used dLUC or iLUC:

- Firstly, aLULUC is based on available data and does not require complex economic models or value-based choices of crucial parameters such as time horizons. This makes results more robust.
- Secondly, a comprehensive and regularly updated database is available based on the respective national inventory reports and FAOSTAT for LUC and LU.
- Thirdly, in contrast to iLUC, aLULUC is compatible to attributional LCA, because it attributes burdens/emissions to products and not to change processes. Finally, aLULUC factors on a country level can help to derive meaningful messages to politicians in charge for protection measures or to consumers.

The LCIA indicator phosphate rock demand was introduced as a stand-alone resource indicator because phosphate is a finite resource that cannot be replaced in its vital major application as fertilizer. Thus, phosphate consumption without recycling needs to be reduced which requires measures that are independent of other finite resources. The phosphate rock footprint was shown to be a valuable tool to identify such measures. For biofuels, for example, hot spots of phosphate use were found in various life cycle stages. This information can easily be lost in common evaluations of common aggregate resource indicators.

In summary, the LCI/LCIA methods aLULUC and phosphate rock demand are suitable to derive additional insights and recommendations for LCAs with a wide range of goals and scopes. In particular, both methods are designed to yield recommendations how to overcome crucial bottlenecks of agriculture that are solution-oriented and useful in practice. Therefore, these methods should be considered as an extension of or, if already addressed, alternative to methods nowadays routinely applied in LCA.

Acknowledgements A joint effort in several projects led to this work. The projects have received funding from the European Union's Horizon 2020 research and innovation programme under grant agreement Nos. 727698 ("Magic"), 763911 ("eForFuel") and 727463 ("BioMates") as well as from the Bio-Based Industries Joint Undertaking (JU) under grant agreement No. 792004 ("UNRAVEL"). The JU receives support from the European Union's Horizon 2020 research and innovation programme and the Bio-Based Industries Consortium. Furthermore, the authors would like to thank all colleagues at ifeu, in particular Guido Reinhardt, Regine Vogt and Nabil Abdalla, for their contributions, discussions and recommendations in developing the LCA methods discussed in this paper.

References

1. Fehrenbach, H., Keller, H., Abdalla, N., & Rettenmaier, N. (2020). *Attributional land use (aLU) and attributional land use change (aLUC) – A new method to address land use and land use change in life cycle assessments, version 2.1 of ifeu paper 03/2018.* Available at www.ifeu. de/en/ifeu-papers/. ifeu – Institute for Energy and Environmental Research Heidelberg.
2. Fehrenbach, H., Grahl, B., Giegrich, J., & Busch, M. (2015). Hemeroby as an impact category indicator for the integration of land use into life cycle (impact) assessment. *International Journal of Life Cycle Assessment, 20*(11), 1511–1527.

3. Lindner, J. P., Fehrenbach, H., Winter, L., Bloemer, J., & Knuepfer, E. (2019). Valuing Biodiversity in Life Cycle Impact Assessment. *Sustainability, 2019*(11), 5628. https://doi.org/10.3390/su11205628

4. Directive (EU) 2018/2001 Of the European Parliament and of the Council of 11 December 2018 on the promotion of the use of energy from renewable sources (recast), *Official Journal of the European Union*, L 328/82.

5. Brunner, P. H. (2010). Substance flow analysis as a decision support tool for phosphorus management. *Journal of Industrial Ecology, 14*(6), 870–873.

6. Killiches, F. (2013). *Phosphat – Mineralischer Rohstoff und unverzichtbarer Nährstoff für die Ernährungssicherheit weltweit*. Bundesanstalt für Geowissenschaften und Rohstoffe, Hannover, Germany.

7. USGS. (2008). Mineral commodity summaries 2008. In *U.S. Geological survey, mineral commodity summaries*. U.S. Geological Survey (USGS), Reston, VA.

8. Gwosdz, W., Röhling, S., & Lorenz, W. (2006). Bewertungskriterien für Industrieminerale, Steine und Erden. Geologisches Jahrbuch 12/2006, Reihe H, Wirtschaftsgeologie, Berichte zur Rohstoffwirtschaft Hannover, Germany.

9. Pinnekamp, J., Everding, W., Gethke, K., Montag, D., Winfurtner, K., Sartorius, C., Von Horn, J., Tettenborn, F., Gäth, S., Waida, C., Fehrenbach, H., Reinhardt, J. (2011): Phosphorrecycling – Ökologische und wirtschaftliche Bewertung verschiedener Verfahren und Entwicklung eines strategischen Verwertungskonzepts für Deutschland.

10. Spörri, A., Erny, I., Hermann, L., & Hermann, R. (2017). *Beurteilung von Technologien zur Phosphor-Rückgewinnung*. Ernst Basler + Partner AG.

11. Cordell, D., Drangert, J.-O., & White, S. (2009). The story of phosphorus: Global food security and food for thought. *Global Environmental Change, 19*(2), 292–305.

12. van Kauwenbergh, S. (2010). World Phosphate Rock. In *Technical Bulletin IFDC*. International Fertilizer Development Center (IFDC).

13. Vaccari, D. A., & Strigul, N. (2011). Extrapolating phosphorus production to estimate resource reserves. *Chemosphere, 84*(6), 792–797.

14. van Vuuren, D. P., Bouwman, A. F., & Beusen, A. H. W. (2010). Phosphorus demand for the 1970–2100 period: A scenario analysis of resource depletion. *Global Environmental Change, 20*(3), 428–439.

15. United Nations. (2017). *World population prospects: The 2017 revision, key findings and advance tables* (Working paper no. ESA/P/WP/248). United Nations, Department of Economic and Social Affairs, Population Division.

16. Reinhardt, G., Rettenmaier, H., & Vogt, R. (2019). *Establishment of the indicator for accounting of the resource "phosphate" in environmental assessments*. ifeu papers 01/2019, available at www.ifeu.de/en/ifeu-papers/. ifeu – Institute for Energy and Environmental Research Heidelberg.

17. Patyk, A., & Reinhardt, G. A. (1997). *Düngemittel – Energie- und Stoffstrombilanzen*. Friedr. Vieweg & Sohn Verlagsgesellschaft mbH.

18. Fehrenbach, H. (2014). ILUC und Nachhaltigkeitszertifizierung - (Un-)Vereinbarkeit, bleibende Lücken, Chancen. [ILUC and sustainability certification - (in)compatibility, remaining gaps, opportunities.] In: *Biokraftstoffe zwischen Sackgasse und Energiewende - Sozial-ökologische und tansnationale Perspektiven*, oekom Verlag, Munich.

19. Finkbeiner, M. (2013). *Indirekte Landnutzungsänderungen in Ökobilanzen - wissenschaftliche Belastbarkeit und Übereinstimmung mit internationalen Standards. [Indirect land use change in life cycle assessments - scientific robustness and consistency with international standards.]* Study commissioned by OVID and UDB, Berlin.

Open Access This chapter is licensed under the terms of the Creative Commons Attribution 4.0 International License (http://creativecommons.org/licenses/by/4.0/), which permits use, sharing, adaptation, distribution and reproduction in any medium or format, as long as you give appropriate credit to the original author(s) and the source, provide a link to the Creative Commons license and indicate if changes were made.

The images or other third party material in this chapter are included in the chapter's Creative Commons license, unless indicated otherwise in a credit line to the material. If material is not included in the chapter's Creative Commons license and your intended use is not permitted by statutory regulation or exceeds the permitted use, you will need to obtain permission directly from the copyright holder.

The Environmental Assessment of Biomass Waste Conversion to Sustainable Energy in the Agricultural Biogas Plant

Magdalena Muradin

Abstract Operating an agricultural biogas plants offers the potential of stable, clean, renewable and diversified energy source. It is also a good opportunity to reduce the amount of organic waste. The objective of this study is to evaluate the main environmental hot spots of operating agricultural biogas plants using LCA methodology. This article presents the environmental impact assessment of two agricultural biogas plants with different type of feedstock provision. The environmental life cycle assessment was carried out from "cradle to gate" using the SimaPro software and the ILCD 2011 Midpoint+ methodology. The boundaries of the system included cultivation of maize, delivery of feedstock to the plant, energy production, storage and transport of digestate. The results show that transport of liquid manure induces the highest environmental impact.

1 Introduction

In 2019, the European Parliament assigned the resolution on the climate and environment emergency. Based on that, it is an urgent need to implement and develop many new technologies especially in energy sector, to prevent the further intensification of the crisis and reduce the global temperature growth.

It was expected that carbon dioxide produced by human activity would be absorbed by the oceans. Meanwhile, by warming the atmosphere, CO_2 is additionally released from the oceans and melting ice, so that its concentration may increase exponentially and cause more and more negative climatic phenomena. Food production and consumption account for as much as 35% of all greenhouse gases in the atmosphere, of which agriculture alone accounts for 10%.

Developed countries are struggling with ever-increasing amounts of waste, including the agri-food industry waste, due to overproduction and consumption of food. The issue of the generation of biodegradable waste is often marginalized, while animal production and the generated livestock manure contribute to 30% of

M. Muradin (✉)
Poznan University of Economics and Business, Poznan, Poland
e-mail: magdalena.muradin@ue.poznan.pl

© The Author(s) 2022
Z. S. Klos et al. (eds.), *Towards a Sustainable Future - Life Cycle Management*,
https://doi.org/10.1007/978-3-030-77127-0_12

the total emission of anthropogenic methane to the atmosphere. The global warming potential for methane is from 23, which means that the same amount of methane in the atmosphere as CO_2 will have a 23 more significant impact on climate warming. Technologies based on anaerobic digestion are very useful in reducing the amount of waste from agri-food industry and at the same time enable controlled methane capture and energy production in cogeneration systems.

Manure is a livestock residue that has little commercial value [1]. The slurry digestate which is a result of the anaerobic fermentation can be used as more bio-available fertilizer form [2] and helps to reduce the number of pathogens entering the soil with direct application. Furthermore, storing animal manure in the open air results in methane and carbon dioxide emissions through the process of self-remediation [3]. Anaerobic digestion of animal manure reduces the environmental impacts caused by carbon dioxide, methane and nitrous oxide emissions from storage and reduces waste and odours [4]. For example, in Finland case, anaerobic fermentation on cattle farms contributes to the reduction at approximately 9% of the national agricultural GHG emission reduction goal during the 2005–2020 period [5].

However, animal manure has low biogas yield (9–36 m^3/Mg) compared with different feedstock especially maize silage. In this case, the co-digestion of different biodegradable substrates is often used at farms. The most effective in producing biogas is digesting liquid manure with maize silage, what is however economically unfavourable, and what even worse, maize cultivation for the energy production purpose stays against cultivation for feeding. The solution could be co-digestion with different waste from agri-food production. Such products have often relatively higher methane yield than manure, e.g. potato pulp or fruit pomace. Very favourable to use as a feedstock is also distillery waste. The methane yield for that waste is lower and similar to liquid manure but to manage with this waste is also very problematic and biogas plant can be a solution.

Anaerobic digestion seems to be a very efficient way to close the material and nutrient loop according to EU circular economy paradigm. The field application of digestate is also a part of nutrients' circularity. Digestated materials have advantages for their use as soil amendments which are microbial stability, hygiene and high amount of N present as ammonium. It improves also the total organic C concentration in soil [6, 7].

Biogas is a promising substitute for natural gas of fossil origin [8]. Published articles about environmental impact of biogas production analysed heat and energy production [9], biomethane purification [10] and domestic use [11]. Reviews also describe biogas LCA from manure in a global perspective, technological studies of biogas production and specific studies for specific countries or region. Studies also concern a transport of feedstock and indicate that it can play an important role in the environmental performance of biogas production [12]. The maximum transportation distance should not extend 10 km to make biogas environmentally viable for small-scale plants [9, 13], and for large-scale plants, it should be within 64 km [14]. However, mostly studies focus on electricity generation from biogas than on the possibility of biomass waste treatment.

The aim of this paper is to present the results of selected two biogas plants, which mostly differ with the type of feedstock, the way it is transported and the transportation distance, in order to highlight the most critical factors (hot spots) from the environmental point of view of operating those installations whose main purpose is the waste treatment.

2   Materials and Methods

The life cycle assessment (LCA) methodology was chosen for this study based on ISO 14040 and ISO 14044 as the most comprehensive evaluation of environmental impact. LCA analysis includes four steps: goal and scope definition, life cycle inventory analysis, life cycle impact assessment and interpretation of results [15, 16]. In this work, the ILCD 2011 Midpoint+ v.1.10 method was considered. The ILCD was developed by the Institute for Environment and Sustainability in the European Commission Joint Research Centre (JRC), in cooperation with the Environment DG which is widely used in Europe. In this method, 16 very detailed impact assessment categories are distinguished [17]. The inventory data for this study were taken directly from tested agricultural biogas plants located in Poland and from the ecoinvent database v. 3.3 and processed using the SimaPro calculation program.

Selected biogas plants were assessed in details from gate-to-gate perspective. Input data were collected for separate unit processes implemented under the modern mesophilic fermentation technology: maize cultivation, feedstock delivery, energy production and digestate storage and transport. All results were analysed relative to the reference unit, which is named as a functional unit (FU) and defined as "a delivery of 1000 Mg of feedstock designed to biogas conversion". The values of the eco-indicator were presented in impact categories, expressing the value of impact at environmental ecopoints (marked with the Pt symbol).

The allocation cut-off by classification model was used in this study, and the primary production of input of raw materials and pig slurry was allocated to the primary user/producer. It was also considered that the main product is electricity with 100% allocation, but the main purpose of those plants is the biomass waste management. Only the maize cultivation was taken into consideration as a dedicated tillage.

Two agricultural biogas plants A and B were taken into consideration with installed power 1.0 MW and 0.526 MW, respectively. In both cases, slurry digestate is not separated and used as a natural fertilizer on arable fields. The most important parameters of the tested plants are collected in Table 1. The construction and demolition of the biogas plant as well as the production of biomass waste feedstock and digestate application on fields were excluded from the scope of the study. The environmental impacts of the electricity production from biogas based on anaerobic co-digestion of pig slurry, silage maize and different feedstock from agri-food industry were determined (Table 2).

Table 1 The most important parameters of the tested biogas plants

Parameter	Biogas plant A	Biogas plant B
The amount of biogas [m³/year]	4,169,760	1,725,155
The amount of electricity produced [MWh/ year]	786.1	300.7
The amount of heat produced [MWh/ year]	776.9	319.3
The amount of heat used [MWh/ year]	147.0	222.1
The amount of digestate [m³/year]	35,515	19,744
Total efficiency [%]	51	69

Table 2 The feedstock input in relation to annual operations

Biogas plant	Type of feedstock	The amount of feedstock [Mg/year]	Biogas yield [m³/Mg]	Maximum transport distance [km]
A	Pig slurry	14,824.0	232.0	5.0
	Maize silage	21,693.0	36.0	1.0
B	Maize silage	2,025.0	230.0	45.0
	Distillery residues	11,489.7	31.0	Gravity pipeline
	Carrot pomace	1,595.9	76.0	11.3
	Potato pomace	5,919.6	94.0	22.5
	Pig slurry	590.0	9.0	3.8
	Protein sediments	402.6	700.0	172.5

Liquid animal manure was transported by a farm tractor with a barrel. Maize harvested from the fields was transported to a biogas plant using heavy wheeled transport. The remaining raw materials from the agri-food industry were transported with a trailer or with different types of lorries. Only distillery residues in biogas plant B were delivered by a gravity pipeline.

3 Results and Discussion

The results were estimated by using the ILCD 2011+ method and the 16 midpoint categories. The results were described on two different levels of LCA methodology: characterization and weighting for four-unit processes – maize cultivation, feedstock delivery, energy production and digestate storage and transport. The feedstock delivery includes transport of agri-food residues, maize ensilaging on-site and delivery to the digester. In analysed biogas plants, we can distinguish six types of transport: (1) road transport of pig slurry to the plant, (2) road transport of raw materials to the plant, (3) pipeline transport to the plant, (4) internal transport on-site, (5) maize transport from the field and (6) digestate transport for final use as fertilizer.

The cumulative environmental impact of biogas plant B (1.48 kPt) is significantly lower than that of biogas plant A (42.66 kPt). The liquid feedstock with low organic mass content and biogas yield in plant B are provided by gravity pipeline. In an installation A, a feedstock is delivered by a tractor with a barrel (Table 3). The highest environmental burdens of biogas plant A stem from the delivery of a feedstock, whereas of plant B, it is related to the storage and transport of the digestate.

The highest value of eco-indicator for plant B concerns the digestate storage and transport. In this case, the transport of liquid digestate takes place using a tractor with a barrel. The storage of the digestate itself does not involve any energy consumption or emissions to the atmosphere. The digestate is stored in a sealed container, so transport in this process is the main contributor. Moreover, the fields for the application of the digestate are located in the vicinity of biogas plant B and the distance is 0.9616 km maximum. The transportation distance of a digestate to biogas plant A is almost twice longer (1,606 km), which can significantly affect the higher environmental impact. The area required for the spreading of the digestate and the maximum transportation distance were calculated as follows [18].

Biogas plant A exhibits a significant impact on almost all categories; however, the contribution of all categories for both biogas plants is almost equal (Table 4).

Only for the water resource depletion category for both biogas plants, the value is below zero. It means that in this category, the environmental impact can be positive. The lower is the value, the more positive is the impact. The reason for obtaining such results for this impact category is the temporary storage of the liquid digestate. It may provide a reservoir of water for field irrigation just next to the fertilization purpose.

Based on the results, three leading groups of factors with the highest environmental impact were separated: transport, electricity consumption and others. The factors were classified in terms of the value of environmental burdens and significance for the impact on climate change. In both biogas plants, transport is the main contributor and represents 99.9% and 98.1% of the total cumulative impact value, respectively (Fig. 1). Even electricity consumption, which in both cases comes from the grid, represents a negligible part of the total impact, 0.03% and 1.56%, respectively (Fig. 1).

Comparing all different types of transport as it was mentioned earlier in this article, for biogas plant A, the highest environmental impact is related to the transport of pig slurry (91%), while for biogas plant B, the impact mainly stems from the transport of the digestate (87%) (Fig. 2). The transport of raw materials to biogas plant B is characterized by a relatively low environmental impact, even though the distance from the production site to the biogas plant is up to 100 km. Raw materials from agri-food industry such as fruits and vegetable pomace have a higher organic mass content and a higher biogas yield per unit weight (Table 2). Then the

Table 3 The cumulative eco-indicator values for individual stages of operation

Biogas plant	Maize cultivation [kPt]	Feedstock delivery [kPt]	Energy production [kPt]	Digestate storage and transport [kPt]
A	0.58	38.68	0.01	3.39
B	0.01	0.18	0.02	1.27

Table 4 LCIA results of each biogas plant on the characterization level

Impact category	Unit	Plant A	Plant B
Climate change	kg CO₂ eq	1.93E+07	957E+05
Ozone depletion	kg CFC − 11 eq	2.30E+00	134E-01
Human toxicity, non-cancer effects	CTUh	5.82E+01	1.85E+00
Human toxicity, cancer effects	CTUh	2.12E+00	7.84E-02
Particulate matter	kg PM2.5 eq	1.82E+04	7.31E+02
Ionizing radiation HH	kBq U235 eq	1.26E+06	6.50E+04
Ionizing radiation E (interim)	CTUe	6.65E+00	3.67E-01
Photochemical ozone formation	kg NMVOC eq	1.59E+05	6.70E+03
Acidification	molc H+ eq	1.61E+05	6.86E+03
Terrestrial eutrophication	molc N eq	5.56E+05	2.36E+04
Freshwater eutrophication	kg P eq	5.93E+03	2.47E+02
Marine eutrophication	kg N eq	5.11E+04	2.17E+03
Freshwater ecotoxicity	CTUe	3.02E+08	1.14E+07
Land use	kg C deficit	1.35E+08	5.37E+06
Water resource depletion	m³ water eq	−1.94E+06	−5.85E+04
Mineral, fossil and ren resource depletion	kg Sb eq	3.27E+03	1.27E+02

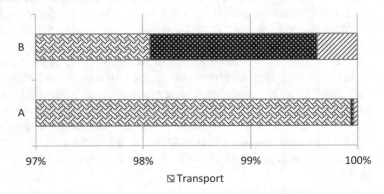

Fig. 1 The contribution of main critical factors in total environmental impact

transportation can be significantly extended obtaining the same results compared with the distance for pig slurry. In plant B, the distillery residues were transported by gravity pipeline what leads to negligible environmental impact at the exploitation stage. The impact can be visible at the construction or demolition stage when the input of metal used for pipelines is taken into account. However, in this study, these two stages were omitted.

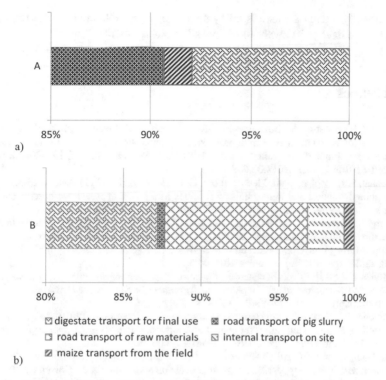

a)

b)

Fig. 2 The share of different types of transport in feedstock delivery process: (a) biogas plant A, (b) biogas plant B

4 Conclusions

The production of renewable energy from biogas is an unquestionably effective way to replace energy from conventional sources and reduce negative environmental impact and climate change. Biogas plants can also provide a solution to the problem of managing many agricultural and agri-food industry waste. However, taking into account many previous studies and this work, we can conclude that transport is the main contributor of the cumulative environmental impact of operating an agricultural biogas plant. Liquid raw materials with a low biogas yield should be transported by pipelines, and biogas plants using this type of raw materials should be located in the vicinity of feedstock sources. This is also confirmed by Cherubini et al. who claimed that keeping animals close to biogas plant provide the reduction of the environmental impact [19].

In the case of the ferment, the distance over which it is to be extracted should be limited or other solutions should be used to limit the quantity of the ferment that is needed to be used, e.g. by drying. The environmental impact of drying processes and the possible pelletization of the resulting biomass should be studied.

Undoubtedly, biomass waste is a key source of renewable energy (not only a biogas), but we have to be aware of the possible environmental impact.

References

1. Battini, F., Agostini, A., Boulamanti, A. K., Giuntoli, J., & Amaducci, S. (2014). Mitigating the environmental impacts of milk production via anaerobic digestion of manure: Case study of a dairy farm in the Po Valley. *Science of the Total Environment, 481*, 196–208. https://doi.org/10.1016/j.scitotenv.2014.02.038
2. Neshat, S. A., Mohammadi, M., Najafpour, G. D., & Pooya, L. (2017). Anaerobic co-digestion of animal manures and lignocellulosic residues as a potent approach for sustainable biogas production. *Renewable and Sustainable Energy Reviews, 79*, 308–322.
3. Burg, V., Bowman, G., Haubensak, M., Baier, U., & Thees, O. (2018). Valorization of an untapped resource: Energy and greenhouse gas emissions benefits of converting manure to biogas through anaerobic digestion. *Resources, Conservation and Recycling, 136*, 53–62. https://doi.org/10.1016/j.resconrec.2018.04.004
4. Möller, K. (2015). Effects of anaerobic digestion on soil carbon and nitrogen turnover, N emissions, and soil biological activity. A review. *Agronomy for Sustainable Development, 35*, 1021. https://doi.org/10.1007/s13593-015-0284-3
5. Timonen, K., Sinkko, T., Luostarinen, S., Tampio, E., & Joensuu, K. (2019). LCA of anaerobic digestion: Emission allocation for energy and digestate. *Journal of Cleaner Production, 235*, 1567–1579. https://doi.org/10.1016/j.jclepro.2019.06.085
6. Al Seadi, T. (2002). Quality management of AD residues from biogas production. In *IEA bioenergy, task 24 – Energy from biological conversion of organic waste*. University of Southern Denmark. http://213.229.136.11/bases/ainia_probiogas.nsf/0/70996A6A88900B70C125753F005B70AD/$FILE/IEA%20BUENAS%20PR%C3%81CTICAS%20DA.pdf. Accessed online on 4 Jun 2021
7. Alburquerque, J. A., de la Fuente, C., & Bernal, M. P. (2012). Chemical properties of anaerobic digestates affecting C and N dynamics in amended soils, agriculture. *Ecosystems & Environment, 160*, 15–22. https://doi.org/10.1016/j.agee.2011.03.007
8. Morero, B., Groppelli, E., & Campanella, E. A. (2015). Life cycle assessment of biomethane use in Argentina. *Bioresource Technology, 182*, 208–216. https://doi.org/10.1016/j.biortech.2015.01.077
9. Boulamanti, A. K., Maglio, S. D., Giuntoli, J., & Agostini, A. (2013). Influence of different practices on biogas sustainability. *Biomass and Bioenergy, 53*, 149–161. https://doi.org/10.1016/j.biombioe.2013.02.020
10. Agostini, A., Battini, F., Giuntoli, J., Tabaglio, V., Padella, M., Baxter, D., Marelli, L., & Amaducci, S. (2015). Environmentally sustainable biogas? The key role of manure co-digestion with energy crops. *Energies, 8*, 5234–5265.
11. Russo, V., & von Blottnitz, H. (2017). Potentialities of biogas installation in South African meat value chain for environmental impacts reduction. *Journal of Cleaner Production, 153*, 465–473. https://doi.org/10.1016/j.jclepro.2016.11.133
12. Hamelin, L., Naroznov, I., & Wenzel, H. (2014). Environmental consequences of different carbon alternatives for increased manure-based biogas. *Applied Energy, 114*, 774–782. https://doi.org/10.1016/j.apenergy.2013.09.033
13. Fantin, V., Giuliano, A., Manfredi, M., Ottaviano, G., Stefanova, M., & Masoni, P. (2015). Environmental assessment of electricity generation from an Italian anaerobic digestion plant. *Biomass and Bioenergy, 83*, 422–435. https://doi.org/10.1016/j.biombioe.2015.10.015

14. Poeschl, M., Ward, S., & Owende, P. (2010). Prospects for expanded utilization of biogas in Germany. *Renewable and Sustainable Energy Reviews, 14*(7), 1782–1797. https://doi.org/10.1016/j rser.2010.04.010
15. ISO 14040:2006 Environmental management – Life cycle assessment – Principles and framework.
16. ISO 14044:2006 Environmental management – Life cycle assessment – Requirements and guidelines.
17. European Commission -Joint Research Centre -Institute for Environment and Sustainability. (2010). *International Reference Life Cycle Data System (ILCD) Handbook -general guide for life cycle assessment -detailed guidance.* First edition March 2010. EUR 24708 EN. Luxembourg. Publications Office of the European Union. Accessed 15 Jan 2020.
18. Hartmann, J. K. (2006). *Life-cycle-assessment of industrial scale biogas plants.* Department for Agricultural Science, Georg-August-Universitat Gottingen. Accessed 15 Jan 2020.
19. Cherubini, E., Zanghelini, G. M., Alvarenga, R. A. F., Franco, D., & Soares, S. R. (2015). Life cycle assessment of swine production in Brazil: A comparison of four manure management systems. *Journal of Cleaner Production, 87*, 68–77. https://doi.org/10.1016/j.jclepro.2014.10.035

Open Access This chapter is licensed under the terms of the Creative Commons Attribution 4.0 International License (http://creativecommons.org/licenses/by/4.0/), which permits use, sharing, adaptation, distribution and reproduction in any medium or format, as long as you give appropriate credit to the original author(s) and the source, provide a link to the Creative Commons license and indicate if changes were made.

The images or other third party material in this chapter are included in the chapter's Creative Commons license, unless indicated otherwise in a credit line to the material. If material is not included in the chapter's Creative Commons license and your intended use is not permitted by statutory regulation or exceeds the permitted use, you will need to obtain permission directly from the copyright holder.

Life Cycle Assessment Benchmark for Wooden Buildings in Europe

Erwin M. Schau, Eva Prelovšek Niemelä, Aarne Johannes Niemelä,
Tatiana Abaurre Alencar Gavric, and Iztok Šušteršič

Abstract Climate change and other environmental problems from the production of raw materials, construction, and end of life of buildings are serious concerns that need to be solved urgently. Life cycle assessment (LCA) and the EU-recommended Environmental Footprint (EF) are well-known and accepted tools to measure a comprehensive set of environmental impacts throughout a product's life cycle. But to assess how good (or bad) a wooden building performs environmentally is still a challenge. In the EU Environmental Footprint [11] pilot phase from 2013 to 2018, an average benchmark for the different product groups was found to be very useful. Based upon the recommendations for a benchmark of all kinds of European dwellings, we developed a scenario of a typical European wooden building. The EU Environmental Footprint method covers 16 recommended impact categories and can be normalized and weighted into one single point for easy and quick comparisons. The results are presented as the average impact per one square meter (m^2) of floor area over 1 year. The developed benchmark for wooden buildings is a suitable comparison point for new wooden building designs. The benchmark can be used by architects and designers early in the planning stages when changes can still be made to improve the environmental performance of wooden buildings or the communication and interpretation of LCA results for customers and other stakeholders.

1 Introduction

According to the European Commission, the construction industry accounts for 15% of all greenhouse gas emissions [1]. During their use phase, buildings use 80% of the total energy consumption [2], which contributes significantly to air pollution and other environmental impacts stemming from energy sourcing, distribution, and transformation. While energy consumption during the use phase is predicted to decrease as efficient buildings, like zero and near zero energy buildings, become more common, climate change and other environmental problems from the production of raw materials, construction, and end of life remain serious concerns that

E. M. Schau (✉) · E. P. Niemelä · A. J. Niemelä · T. A. Alencar Gavric · I. Šušteršič
InnoRenew Centre of Excellence (CoE), Izola - Isola, Slovenia
e-mail: erwin.schau@innorenew.eu

© The Author(s) 2022
Z. S. Klos et al. (eds.), *Towards a Sustainable Future - Life Cycle Management*,
https://doi.org/10.1007/978-3-030-77127-0_13

need to be solved urgently. This calls for a life cycle-based approach for the assessment of the environmental impacts of a building.

In the EU Environmental Footprint [11] pilot phase from 2013 to 2018, an average benchmark for different product groups was found to be very useful [3–5] as a help for interpretation of the product's life cycle assessment results in scope of the product category.

Spirinckx et al. [6] give recommendations on benchmarks for office buildings, while Lavagna et al. [2] provide the average environmental impacts of existing dwellings in Europe. However, as the European Union has introduced a stricter policy for buildings' use of energy, a benchmark for new buildings to be built is needed. In this work, we provide an environmental benchmark for a near zero energy wooden residential buildings (nZEB) for new buildings in the future (after 2020). The typical (European average) wooden single-family house holds on average 2.36 inhabitants and, in this study, is set to be 100 m² large.

2 Data and Method

2.1 Background Data for a Typical (European Average) Wooden Single-Family House

Based on market-based statistics from Eurostat [7], supplemented with national data where necessarily [8], a prevision for where wood-based residential housing is found in Europe today is made (cf Table 1).

The apparent consumption is what is sold in each country and calculated based on production value – export + import (EUR). The apparent consumption is used for weighting the climate data and energy requirement data of the countries investigated to come to an average wooden residential building.

European countries have different climate and, therefore, different heating demand for residential buildings. We took the climatic conditions on a country level into account, represented by the degree heating days, which is a measurement for how much heating is necessary during a year [9, 10]. Table 1 also shows the heating degree days in the countries investigated. The weighted average heating degree days for the European countries according to Table 1 is 3500. We have used 10 years of data for the climate conditions, and not the usual 30 years, for two reasons: (1) prefabricated building statistics are not easily available for 30 years (for weighting the data), and, more importantly, (2) climate is changing to warmer conditions such that an increase in heating degree days can be observed. For example, the reference climate in Germany is 500 heating degree days less (i.e., warmer) in the period 2008–2017 than was used as a reference 20 years ago (3500 heating degree days).

The energy requirements for new residential buildings from 2021 are given in Table 2.

Table 1 Apparent consumption (million EUR) of prefabricated wooden buildings and climate expressed as heating degree days in different countries (average per year, 2008–2017)

Country	Consumption (million EUR)	Heating degree days per year	Country	Consumption (million EUR)	Heating degree days per year
Austria	583	3482	Latvia	5	4046
Belgium	56	2697	Lithuania	65	3854
Bulgaria	5	2494	Luxembourg	7	2906
Croatia	11	2281	Malta	0.1	468
Cyprus	1	691	Netherlands	150	2721
Czechia	27	3309	Norway	544	4113
Denmark	121	3244	Poland	4	3370
Estonia	23	4224	Portugal	14	1201
Finland	414	5466	Romania	30	2924
France	231	2380	Slovakia	10	3173
Germany	1658	3053	Slovenia	25	2785
Greece	2	1546	Spain	143	1742
Hungary	10	2668	Sweden	1126	5221
Ireland	42	2821	United Kingdom	1226	3033
Italy	615	1875	–	–	–

Source: [7–10]

Table 2 Energy requirement for new buildings (nZEB) from 2021

Country	Max kWh/(m² year)	Country	Max kWh/(m² year)	Country	Max kWh/(m² year)
Austria	160.0	Germany	48.3	Norway	97.5
Belgium	45.0	Greece	57.5	Poland	67.5
Bulgaria	40.0	Hungary	61.0	Portugal	57.5
Croatia	37.0	Ireland	45.0	Romania	155.0
Cyprus	100.0	Italy	57.5	Slovakia	43.0
Czechia	57.5	Latvia	95.0	Slovenia	47.5
Denmark	20.0	Lithuania	77.5	Spain	57.5
Estonia	75.0	Luxembourg	57.5	Sweden	52.5
Finland	130.0	Malta	40.0	United Kingdom	44.0
France	52.5	Netherlands	57.5	–	–

Source: Own calculations and estimates based on [12–15]

The weighted average maximum energy requirement (near zero energy building) is 67.5 kWh/(m² year).

2.2 Design of a Typical (European Average) Wooden Single-Family House

With the average climate (from Table 1, 3500 degree heating days, which corresponds to approximate climatic conditions in Austria, South Germany, Slovenia and Italy near the alps) and energy requirement, we started the design of the wooden single-family house that would serve as a benchmark; the shape of the house was made according to the most common plans and structures that we found offered from construction firms of prefabricated wooden houses in Austria. It contains three bedrooms, a living room, cabinet, toilet, utility, staircase, and bathroom. The outer measurements of the house are 9.6 m x 6.7 m, and maximum height is 7.72 m above ground floor level. The house has a pitched roof with 35° angle and 1.0 m overhang. Wooden windows (triple glazed) and doors have Uw = 0.8 W/m²K. There is a 25-cm-thick concrete plate for the house's foundation. Walls are made of wooden profiles 16/8 cm and stone wood filling in-between, with additional 10 cm of stone wool on the outer side covered with finishing plaster. The roof structure is made of 16/8 wooden profiles as well, with mineral wool in-between and 10 cm on top. For roof cover, wave fiber cement roof tiles were used. Inner floors were covered with parquet on floating screed; ceramics were used in sanitary rooms. Figure 1 shows two profiles and Fig. 2 the schematic floor plan of the house.

After preliminary drawings were made, load-bearing construction of the building was calculated and drawings were updated; the layers for all building parts were precisely defined and U-values of the building's outer enclosure were calculated with diverse online tools. Afterward, the house's energy consumption was

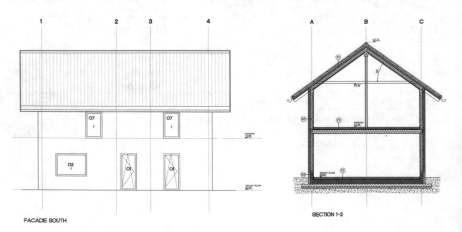

Fig. 1 Façade and section drawings of the house

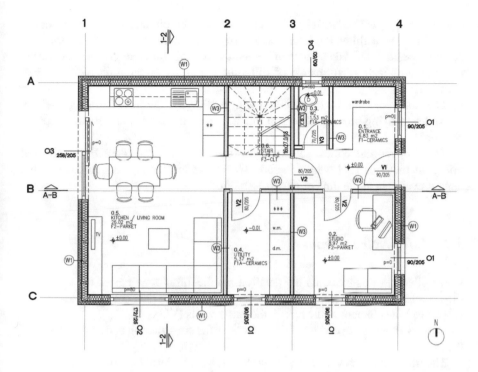

GROUND FLOOR: 50,51 m2

Fig. 2 Ground floor of the house

calculated using a simplified building energy calculation, the Preliminary Passive House Planning Package (PHVP) 2002 [16], which is suitable in the preliminary design phase. Since the shape of the building was made simple and compact, avoiding placement of widows on the northern façade, the energy consumption was calculated to be 26.9 kWh/m²a. This corresponds to nZEB buildings for all countries in Table 2, except for Denmark where there is a stricter requirement.

3 Life Cycle Assessment of a Typical (European Average) Wooden Single-Family House

3.1 Goal and scope

The goal of the life cycle assessment (LCA) for the average wooden one family house is to have a benchmark for wooden buildings suitable as a comparison point for new wooden building designs. The benchmark should be of use for architects and designers early in the planning stages when changes to the building can be made to improve the environmental performance of wooden buildings. Further, a goal of the LCA is to facilitate the interpretation and communication of LCA results for customers and other stakeholders of wooden buildings, for example, when comparing environmental performance of different materials or building elements like the façade.

The functional unit is one dwelling with a 100-year lifetime. Our single-family house has a living area equal to 100 m^2; however, the results are given as per m^2 per year.

The impact categories selected are the EU-recommended Environmental Footprint methods [11], which include 16 impact indicators. Version 2.0 was the newest available at the time of the assessment.

3.2   Life Cycle Inventory

Data collection was based on the detailed architectural drawings of the house (cf. Figs. 1 and 2 for examples). Table 3 shows an example of data collection and calculations for one element of the house, the inner walls (W3).

Table 4 shows an overview of the materials for construction and maintenance of the house.

The life cycle inventory data and modeling follow closely the data and life cycle inventory modeling of the benchmark for environmental impact of housing in Europe – Basket of Products Consumer Footprint indicator for housing [2, 17], where the ecoinvent database is used. We used ecoinvent version 3.5 [18] with allocation, cutoff by classification, as implemented in SimaPro v 9.0 [19] for the background data.

4   Results

The characterized results (cf. Table 5) show that the energy for heating and water use in the operational stage (B6 and B7) of the house is dominating, expect for *land use* and *resource use, minerals, and metals* impact categories, where the product stages (A1–A3), respectively, and maintenance (B2 and B4–B5) are dominating. This is caused by high land use and land transformation for wood products (forest management areas) and high use of materials in the maintenance period, which is quite long (100 years). The *water scarcity* impact category is totally dominated by the operational water use during the use phase. However, both *water scarcity* and *resource use, minerals, and metals* are expected to decrease when the total life cycle, including water and other materials end of life, is included, as these can be cleaned and released into nature or, respectively, become recycled material.

The normalized results in Fig. 3 not only show high *water scarcity* from the use of water in the operational phase but also high *resource use, energy, particulate*

Table 3 Example of data collection, here for inner walls (W3)

W3 – inner walls	Quantity [m²]	Volume [m³]	Mass [kg]
Gypsum plasterboards 1.25 cm*2 = 2.5 cm	92.54	2.313	2082.1
Load-bearing construction profiles 6/10 cm – 10 cm	18.5	1.851	777.3
Stone wool (between wooden construction) – 10 cm	92.5	9.254	277.6
Gypsum plasterboards – 1.25 cm*2 = 2.5 cm	92.54	2.313	2082.1

Table 4 Material quantities for construction and maintenance

Material	Quantities for construction [kg]	Quantities for maintenance [kg]
Concrete	57621	0
Gypsum	9922	17186
Wood	12707	5354
Sawnwood	7419	821
Window frame, wood	1681	3122
Oriented strand board	1502	0
Fiberboard	423	987
Glued laminated timber	1258	0
Door, inner, wood	356	356
Door, outer, wood-glass	67	67
Insulation, stone wool	4355	10161
Cement	4342	2466
Gravel	5858	0
Ceramic	1439	1923
Glass	1019	1892
Plastic	660	806
Steel	1286	41
Insulation, polystyrene	288	673
Glue	395	547
Bitumen	591	0
Copper	23	23
Aluminum	12	0

matter, and *climate change.* Here, the use phase is still important, but so are the product stage (A1–A3) and maintenance (B2 and B4–B5) in these three impact categories.

The weighted results (cf. Figure 4) show that *water scarcity* and *climate change* are the most important, followed by *resource use, energy,* and *respiratory inorganics.* The impact category *ozone depletion* is less relevant.

Table 5 Characterized results [per m^2 and year] broken down at different stages

Impact category (unit)	A1–A3 product stages	A4–A5 transport and construction	B2, B4, B5 maintenance	B6, B7 use – operational energy and water
Climate change (kg CO_2 eq)	2.99E+00	3.90E-01	3.73E+00	8.54E+00
Ozone depletion (kg CFC11 eq)	2.60E-07	5.31E-08	5.83E-07	6.52E-07
Ionizing rad. (kBq U-^{235} eq)	1.42E-01	5.04E-02	1.62E-01	9.40E-01
Photochem. Ozon form. (kg NMVOC eq)	1.24E-02	1.34E-03	1.45E-02	2.68E-02
Respiratory inorg. (disease inc.)	5.35E-07	1.60E-08	6.30E-07	9.76E-07
Non-cancer HH effects (CTUh)	5.09E-07	4.66E-08	5.31E-07	1.77E-06
Cancer HH effects (CTUh)	9.34E-08	3.40E-09	7.56E-08	1.33E-07
Acidification (mol H$^+$ eq)	1.95E-02	2.52E-03	2.71E-02	6.33E-02
Eutrophication – fresh w. (kg P eq)	1.95E-04	2.94E-05	2.26E-04	7.44E-04
Eutrophication – marine (kg N eq)	3.23E-03	4.38E-04	3.56E-03	8.36E-03
Eutrophication terr. (mol N eq)	4.34E-02	6.67E-03	5.24E-02	1.33E-01
Ecotoxicity freshwater (CTUe)	3.01E+00	3.90E-01	3.56E+00	4.21E+00
Land use (Pt)	7.97E+02	4.07E+00	3.55E+02	3.87E+02
Water scarcity (m3 depriv.)	1.06E+00	8.24E-02	1.34E+00	5.73E+01
Resource use, energy (MJ)	4.13E+01	7.11E+00	5.39E+01	1.41E+02
Resource use, mineral, and metals (kg Sb eq)	3.19E-05	7.82E-07	4.29E-05	9.35E-06

5 Discussion, Outlook, and Conclusion

This contribution shows how we designed an average European wooden residential building and used life cycle assessment (LCA) and, more specific, the EU-recommended Environmental Footprint (EF) to investigate the cradle to gate and use phase of the house suitable for a benchmark. Even with an improved design,

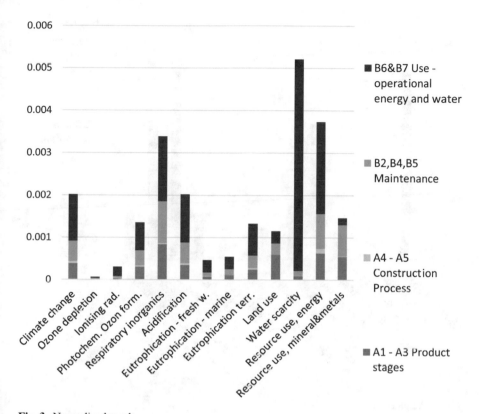

Fig. 3 Normalized results

like better insulation, the use phase is still a major contributor to the environmental impact categories investigated. Climate change, respiratory inorganics (particulate matter), water scarcity, and resource use and energy are the most important impact categories in this study. Waste scenarios, some that happen 100 years into the future, are left for future studies. However, these are believed to include lots of reuse and material recycling. Future studies should also apply the new EU Environmental Footprint method v.3, where the toxicity impact categories have been updated. However, this was not yet implemented in the software used at the time of impact assessment calculation.

The results will be used to compare to existing housing in the Basket of Products for a single-family house and establish and compare the reference houses in specific countries, like Spain. Other building types, like multifamily houses and other buildings made of wood, could be investigated based on the same concept.

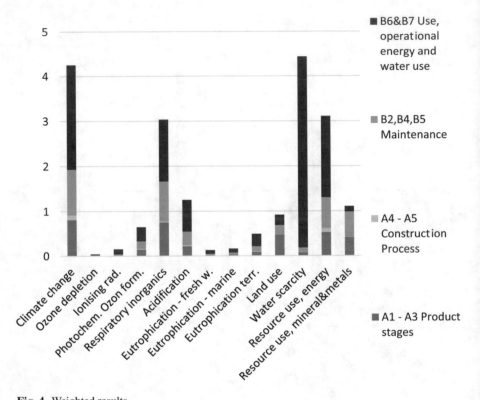

Fig. 4 Weighted results

Acknowledgments The authors gratefully acknowledge the European Commission for funding the InnoRenew CoE project (grant agreement #739574), under the H2020 Widespread-Teaming programme, and Slovenia (investment funding of the Republic of Slovenia and the European Union's European Regional Development Fund).

References

1. European Commission. (2016). Commission recommendation (EU) 2016/1318 of 29 July 2016 on guidelines for the promotion of nearly zero-energy buildings and best practices to ensure that, by 2020, all new buildings are nearly zero-energy buildings. *Official Journal of the European Union*. https://op.europa.eu/s/pdT7
2. Lavagna, M., Baldassarri, C., Campioli, A., Giorgi, S., Dalla, A., Castellani, V., & Sala, S. (2018). Benchmarks for environmental impact of housing in Europe : Definition of archetypes and LCA of the residential building stock. *Building and Environment, 145*(May), 260–275.

3. Schau, E. M. (2019). Product Environmental Footprint (PEF) Category Rules (PEFCR) for intermediate paper products – Overview and discussion of important choices made in the development. In I. Karlovits (Ed.), *Proceedings of the 1st International Conference on Circular Packaging* (pp. 175–184). Pulp and Paper Institute. https://doi.org/10.5281/zenodo.3430522

4. Guiton, M., & Benetto, E. (2018). Special session on product environmental footprint. In E. Benetto, K. Gericke, & M. Guiton (Eds.), *Designing sustainable technologies, products and policies* (pp. 515–520). Springer. https://doi.org/10.1007/978-3-319-66981-6

5. Gül, S., Spielmann, M., Lehmann, A., Eggers, D., Bach, V., & Finkbeiner, M. (2015). Benchmarking and environmental performance classes in life cycle assessment – Development of a procedure for non-leather shoes in the context of the Product Environmental Footprint. *International Journal of Life Cycle Assessment*, 1640–1648. https://doi.org/10.1007/s11367-015-0975-7

6. Spirinckx, C., Thuring, M., Damen, L., Allacker, K., Ramon, D., Mirabella, N., ... Passer, A. (2019). Testing of PEF method to assess the environmental footprint of buildings – Results of PEF4Buildings project. *IOP Conference Series: Earth and Environmental Science*, 297, 012033.

7. Eurostat. (2019). Sold production, exports and imports by PRODCOM list (NACE Rev. 2) – annual data [DS-066341] – Prefabricated buildings of wood.

8. SSB. (2019). *ProdCom 10455: Solgt produksjon av varer for store foretak i industri (In Norwegian: Sold production of goods in the manufacturing industry)*. Statistics Norway.

9. Eurostat. (2019). *Cooling and heating degree days by country – annual data* [nrg_chdd_a].

10. Enova. (2019). Graddagstall (In Norwegian: Degree heating days): Oslo https://www.enova.no/om-enova/drift/graddagstall/. Accessed 06 June 2019.

11. European Commission. (2013). Commission Recommendation of 9 April 2013 on the Use of Common Methods to Measure and Communicate the Life Cycle Environmental Performance of Products and Organisations - 2013/179/EU. L 124: 1–210. *Official Journal of the European Union*. http://data.europa.eu/eli/reco/2013/179/oj

12. BPIE. (2015). *Nearly zero energy building*. Buildings Performance Institute Europe (BPIE). http://bpie.eu/publication/nzeb-definitions-across-europe-2015/. Accessed 28 Mar 2019.

13. D'Agostino, D., & Mazzarella, L. (2019). What is a Nearly zero energy building? Overview, implementation and comparison of definitions. *Journal of Building Engineering, 21*, 200–212. https://doi.org/10.1016/j.jobe.2018.10.019

14. Kurnitski, J., & Ahmed, K. (2018). *NERO – Cost reduction of new nearly-zero energy wooden buildings in Northern climate conditions – D1.2*. Summary report on nZEB requirements.

15. NRW ÖkoZentrum. (2019). *Gesetzentwurf der Bundesregierung (in German: Draft bill from the German government)*. http://www.oekozentrum-nrw.de/fileadmin/Medienablage/PDF-Dokumente/190528_GEG-Entwurf.pdf. Accessed 28 May 2019.

16. Feist, W., Baffia, E., Schnieders, J., & Pfluger, R. (2002). *Energiebilanzverfahren für die Passivhaus Vorprojektierung 2002 (PHVP02)*. Darmstadt. https://passivehouse.com/05_service/02_tools/02_tools.htm. Accessed 22 July 2019.

17. Baldassarri, C., Allacker, K., Reale, F., Castellani, V., & Sala, S. (2017). *Consumer footprint: Basket of products indicator on housing*. Publications Office of the European Union. https://doi.org/10.2760/05316

18. Ecoinvent Centre. (2018). Ecoinvent life cycle inventory database, v 3.5.

19. Pré Consultants. (2019). SimaPro analyst, v. 9.0.

Open Access   This chapter is licensed under the terms of the Creative Commons Attribution 4.0 International License (http://creativecommons.org/licenses/by/4.0/), which permits use, sharing, adaptation, distribution and reproduction in any medium or format, as long as you give appropriate credit to the original author(s) and the source, provide a link to the Creative Commons license and indicate if changes were made.

The images or other third party material in this chapter are included in the chapter's Creative Commons license, unless indicated otherwise in a credit line to the material. If material is not included in the chapter's Creative Commons license and your intended use is not permitted by statutory regulation or exceeds the permitted use, you will need to obtain permission directly from the copyright holder.

Importance of Building Energy Efficiency Towards National and Regional Energy Targets

Can B. Aktaş

Abstract The buildings sector in the EU consumes 40% of energy and is responsible for 36% of CO_2 emissions. With growing public interest on the subject, there have been several EU policies developed to curb impacts. Statistical analysis conducted in the case study indicates an increase in both total and buildings' energy consumption trends leading up to 2030, with total energy consumption having an expected value of 40% increase and building energy consumption having an expected value of 33% increase. Analysis results indicate that building energy consumption could be maintained at current levels if a proactive approach is embraced. Focusing solely on buildings' energy consumption does not solve national or regional energy problems, but neglecting them altogether prevents significant gains to be made. Building energy efficiency is not the solution by itself to achieve energy goals in EU, but is an important contributor toward the solution.

1 Introduction

In the EU, buildings are responsible for approximately 40% of energy consumption, and 36% of CO_2 emissions. Approximately 40% of residential buildings in EU are dated pre-1960, with another 45% from between 1960 and 1990 and did not undergo major renovation since then. Currently, almost 75% of the building stock in the EU is reported to be energy inefficient [1]. Building energy efficiency measures are known to generate economic, societal, and environmental benefits. They also stimulate the economy, in particular the construction industry which generates about 9% of EU's GDP and directly accounts for 18 million jobs. Especially SMEs are known to benefit from building energy efficiency measures as they contribute to more than 70% of the value added in the EU building sector [1].

Existing EU policies demonstrate the timeliness of the subject as successive EU policies regarding building energy efficiency have been put forth in recent years

C. B. Aktaş (✉)
Department of Civil Engineering, TED University, Ankara, Turkey
e-mail: can.aktas@tedu.edu.tr

© The Author(s) 2022
Z. S. Klos et al. (eds.), *Towards a Sustainable Future - Life Cycle Management*,
https://doi.org/10.1007/978-3-030-77127-0_14

including the 2010 Energy Performance of Buildings Directive and the 2012 Energy Efficiency Directive. The former directive has a 2020 strategy of making new construction nearly zero-emission buildings [2]. Hence, there is urgency toward further action as goals are already set to curb energy consumption and associated emissions.

Sandberg et al. [3] demonstrate that the intended EU energy efficiency goals cannot be met if the best available energy efficiency measures are not applied when existing dwellings undergo renovation during their lifetime. While existing building energy codes and regulations are a step forward in the right direction, they have not proven to be sufficient to achieve desired efficiency gains. Furthermore, developers and consumers alike have been shown to interpret meeting the minimum requirements set by the code as sufficient warranty for the energy efficiency of the building, whereas the code rarely represents the optimal point of efficiency [4, 5]. There have been developments in numerous building efficiency technologies to reduce energy consumption in buildings, but their implementation has been lagging mostly due to a lack of knowledge or awareness of their potential impacts, which could be significant considering the extensive lifetime of residential buildings.

The goal of this study is to identify the extent building energy efficiency can play a role toward meeting national and regional energy targets. For that purpose, total energy consumption together with the building sector's share has been analyzed together with forecasts for the near future in line with EU Directives timeline.

Data on Turkey was analyzed as a case in point, as it is one of the fastest growing economies in the EU region as well as having one of the highest total energy demand in the region. Turkey's population grew from 56.5 million in 1990 to 71.5 million in 2008. In addition to population growth, Turkey's urbanization rate has also increased from 52.9% to 74.9% during those years. As a result of these population movements, the number of buildings and consequently energy consumption in buildings increased rapidly [6]. As a result of the developing economy and increasing urbanization rate, electricity consumption has tripled between 1990 and 2008 and reached 198 TWh. Furthermore, Turkey has experienced the highest increase in energy demand in the past 10 years among OECD countries, and only second after China globally. Current expectations are that the trend will continue in short and medium terms [6, 7].

2 Turkey's Total and Sectoral Energy Demand

Between 1972 and 2017, Turkey's total energy consumption rose from 20 million ton petrol equivalent (TPE) to 111 million TPE, indicating a 5.5-fold increase in total energy consumption within 45 years. Figure 1 presents total and sectoral energy consumption trends, both via historical data, as well as forecasted levels of consumption via a statistical analysis that has been carried out. It can be observed that exponential distribution provides the best fit to past data with the indicated R^2 values, as compared to a linear trend [8].

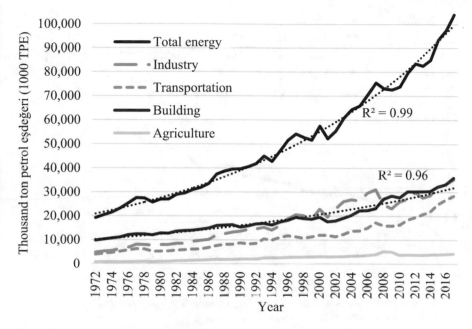

Fig. 1 Total and sectoral energy consumption in Turkey between 1972 and 2017 [8]

Forecasting methods up to the year 2030 have been carried out by using statistical methods. The tool of choice was "Crystal Ball" software. Forecast assessment carried out using the autoregressive integrated moving average (ARIMA) model provided a 95% confidence interval for the expected energy consumption level by 2030. In this context, total consumption and consumption in buildings are presented separately in Figs. 2 and 3 for closer examination of the range, and their implications.

The average value of expected total energy consumption in 2030 is 152 million TPE, and with 95% probability consumption is expected to be between 122 and 182 million TPE. The average value indicates an increase of 40% should be expected compared to 2017 levels. Considering the confidence interval, an increase of 10–65% may be expected by 2030 with a probability of 95%. What should also be emphasized is that it is very unlikely that total energy consumption will remain constant, let alone decrease, in the next decade in Turkey [8].

The average value of forecasted building energy consumption is 48 million TPE for 2030. The 95% confidence interval indicates that consumption may be expected to be in between 37 and 59 million TPE. These values indicate that the average consumption will increase by 33% from the 36 million TPE level in 2017, will remain flat in the best-case scenario, and will increase by 64% in case of a rapid increase.

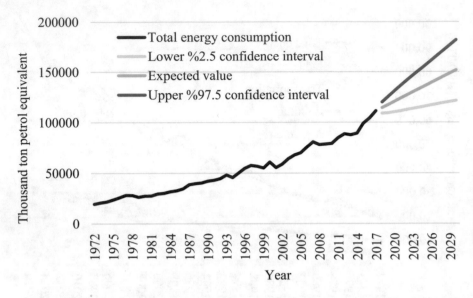

Fig. 2 Average estimated value of the total energy consumption forecasted for 2030 together with its 95% confidence interval [8]

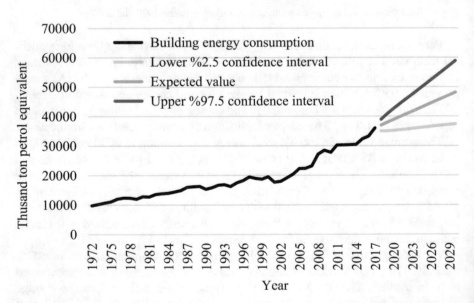

Fig. 3 Average estimated value of building energy consumption forecasted for 2030 together with its 95% confidence interval [8]

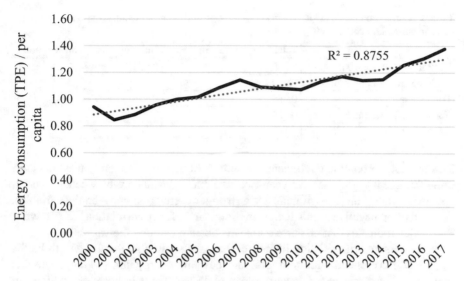

Fig. 4 Energy consumption per capita in Turkey between 2000 and 2017 in TPE [9–10]

2.1   Energy Consumption Per Capita

Energy consumption per capita is important both for forecasting energy consumption levels and for comparisons among countries. For this reason, per capita energy consumption was analyzed and presented in Fig. 4. Since annual population information could not be obtained from reliable sources before 2000, the evaluation was limited to 2000–2017. Results indicate that in addition to an increase in population in Turkey, per capita energy consumption has also increased consistently together, leading to an even more rapid increase in total energy consumption. This trend overlaps with those seen in other countries in the EU and elsewhere.

2.2   Factors that Contribute to Total and Building Energy Consumption

The most frequently studied factor looking into the causality of energy consumption of countries is economic activity or gross domestic product (GDP). There are several detailed studies on the subject in academic literature [11–12]. However, as part of the case study, it was deemed valuable to not only analyze GDP but also investigate the correlation between energy consumption and other pertinent factors. Among the factors analyzed were factors such as population, foreign exchange rate index, and oil price index.

The correlation between the above listed factors with energy consumption was evaluated using the Pearson correlation coefficient and results presented in Table 1.

Table 1 Correlation between total energy consumption and analyzed factors. Presented values are Pearson correlation coefficients [13–16]

	Total energy consumption	Building energy consumption
GDP index (1972 = 100)	0.989	0.985
Building energy consumption	0.979	–
Population	0.968	0.976
Foreign exchange index	0.924	0.912
Oil price index	0.747	0.780

The Pearson correlation coefficient is a statistical method frequently used to determine the linear correlation between two variables. Results are between −1 and 1, and as the values increase, it indicates a stronger correlation between the variables. A positive or negative result indicates direct or indirect correlation, respectively. Zero value indicates that no correlation was detected among the variables.

In agreement with existing literature, the case study also found GDP to be the main factor correlated with both total and building energy consumption. The fact that this analysis was based on a time span of 45 years may indicate that policies or studies aiming to forecast future energy consumption should pay close attention to GDP. Another outcome of the analysis is the revelation on the close correlation between total and building energy use. At least in the past 45 years, the two can be said to have moved together.

3   Role of Building Energy Efficiency Targets Towards National Goals

In order to maintain the current national energy consumption level, energy efficiency policies will need to be developed, enacted, and regulated in order to minimize a further increase in energy consumption. As was discussed in Sect. 2, the expected value of total energy consumption in 2030 is 152 million TPE with a 95% confidence interval of 122–182 million TPE. For buildings, the expected value was 48 million TPE with a 95% confidence interval of 37–59 million TPE.

The abovementioned statistical values were taken as a basis in determining the energy efficiency targets required for the residences to stabilize or reduce the national energy consumption. In this context, the aim is to reduce the energy consumption level as much as possible with effective policies and techniques. Existing data and assessment of Turkey's total energy levels to maintain the level of 2017 indicate that this goal is not achievable only through improving the energy efficiency of buildings. The expected increase in total energy consumption of 41 million TPE is higher than the entire energy consumed by buildings in 2017. Therefore, it seems unlikely that total energy consumption will stabilize or decrease in the short term. Increasing population and per capita energy consumption values also support this result. What needs to be done is to establish and implement effective policies toward these targets with the assumption that the environmental, social, and economic goals and priorities will be determined, and energy consumption will increase. While setting official goals and targets aiming for stabilizing or reducing total national energy consumption, statistical analysis of past policies and practices of the past 45 years indicates that such goals may be beyond reach at least for certain countries. They may still have motivational value, but lack a strong scientific basis unless drastic technological changes are mandated and implemented.

However, when energy consumption of buildings is examined, it seems possible that the increase may be reversed with a proactive approach. The stated expected value assumes that the methods and techniques applied to date will continue to change at the same rate moving forward. However, increasing energy consumption in buildings can be prevented with effective policies and methods. The numerical target determined for this purpose is an additional 25% energy efficiency in buildings based on their current state of energy consumption. However, this strategy should be applied not only to new buildings but also to existing ones, as failure to improve the performance of existing buildings mostly negates any significant gains that may be achieved through new buildings alone. It is not possible to reach the desired energy consumption target set by the EU Directive on buildings by 2030 with policies targeting only the construction of new buildings. Ultimately, even though focusing solely on buildings' energy consumption do not solve national or regional energy problems, due to the share of energy consumed in buildings, neglecting them altogether prevents significant gains to be made. Therefore, building energy efficiency is not the solution by itself to achieve regional energy goals, but is an important contributor toward the solution.

The analysis described herein was based on a case study of Turkey. The reasons for its selection were explained previously and include the fact that Turkey is one of the fastest growing economies in the region and has one of the highest energy demands. However, the conclusions from the analysis do not stay limited to one country, and similar results may be expected for the EU region in general as the underlying principles and factors that affect energy consumption remain the same. Therefore, the presented case study sheds light on the influence and potential impact of building energy use toward national and regional energy goals.

4   Conclusions

The buildings sector in the EU is significant when dealing with energy or environmental issues as buildings consume 40% of energy and are responsible for a comparable amount of greenhouse gas emissions. On the other hand, 85% of buildings in the EU are built before 1990, with 40% built before 1960. This is a problem as well as an advantage: energy-inefficient homes have led to higher than required energy consumption in the EU region; but potential gains to be made by employing efficiency measures are significant. With growing public interest on the subject, there have been several EU projects, guidelines, and policies developed to curb energy consumption and associated emissions.

Turkey is used as a case study in this study as the country has one of the fastest growing economies in the region, and also has a high energy demand growth, which is the central theme of the study. Both total energy consumption and building energy consumption in Turkey have increased exponentially in the past 45 years, although building energy consumption could possibly be represented by a linear trendline as well. This is positive as it indicates a certain degree of energy efficiency measures taking hold in the buildings sector.

Statistical analysis conducted in the analyzed case study indicates an increase in both total and buildings' energy consumption trends leading up to 2030, with total energy consumption having an expected value of 40% increase with a 95% confidence interval of 10–65%, and building energy consumption having an expected value of 33% and a 95% confidence interval of 3–64%. Analysis results indicate that total energy consumption should be expected to increase even in the best-case scenario, but building energy consumption could be maintained at current levels if a proactive approach is embraced.

Multiple factors were analyzed to test correlation with energy consumption. Among the variables analyzed, GDP was found to be highly correlated with energy consumption both for total and for building energy consumption with a Pearson correlation coefficient of 0.99 for both. This fact could provide a quick way of estimating future changes in energy consumption in other countries and regions as well.

Results of the study indicate that it is not possible to reach the desired energy consumption target set by the EU Directive on buildings by 2030 with policies targeting only the construction of new buildings. Ultimately, even though focusing solely on buildings' energy consumption does not by themselves solve national or regional energy problems, due to the share of energy consumed in buildings, neglecting them altogether prevents significant gains to be made. Therefore, building energy efficiency is not the solution by itself to achieve regional energy goals, but is an important contributor toward the solution.

References

1. European Commission – Buildings. https://ec.europa.eu/energy/en/topics/energy-efficiency/buildings
2. Energy Performance of Buildings Directive, Directive 2010/31/EU of the European Parliament and of the Council of 19 May 2010 on the energy performance of buildings, https://eur-lex.europa.eu/legal-content/EN/TXT/HTML/?uri=LEGISSUM:en0021&from=EN&isLegissum=true
3. Sandberg, N. H., Sartori, I., Heidrich, O., et al. (2016). Dynamic building stock modelling: Application to 11 European countries to support the energy efficiency and retrofit ambitions of the EU. *Energy and Buildings, 132*, 26–38. https://doi.org/10.1016/j.enbuild.2016.05.100
4. Laustsen, J. (2008). *Energy efficiency requirements in building codes, energy efficiency policies for new buildings*. International Energy Agency.

5. Morrissey, J., & Horne, R. E. (2011). Life cycle cost implications of energy efficiency measures in new residential buildings. *Energy and Buildings, 43*(4), 915–924. https://doi.org/10.1016/j.enbuild.2010.12.013
6. UNDP. (2010). *Promoting energy efficiency in buildings*. United Nations Development Programme.
7. MFA. (2017). *Turkey's energy profile and strategy*. Ministry of Foreign Affairs, http://www.mfa.gov.tr/turkeys-energy-strategy.en.mfa. Accessed 01 June 2018.
8. Aktaş, C. B. (2019). Ulusal enerji tüketiminin değerlendirmesi ve istatistiksel tahmini. *Bitlis Eren Üniversitesi Fen Bilimleri Dergisi, 8*(4), 1422–1431. https://doi.org/10.17798/bitlisfen.542963
9. Enerji İşleri Genel Müdürlüğü. *İstatistikler – Denge Tabloları*. http://www.eigm.gov.tr/tr-TR/Denge-Tablolari/Denge-Tablolari?page=1. T.C. Enerji ve Tabii Kaynaklar Bakanlığı. Accessed 01 July 2018.
10. TÜİK. *Temel İstatistikler, Nüfus ve Demografi – Nüfus İstatistikleri – Yıllara Göre İl Nüfusları 2000–2018*. http://www.tuik.gov.tr/UstMenu.do?metod=temelist. Türkiye İstatistik Kurumu. Accessed 01 Feb 2019.
11. Korkmaz, Ö., & Develi, A. (2012). Türkiye'de Birincil Enerji Kullanımı, Üretimi ve Gayri Safi Yurt İçi Hasıla (GSYİH) Arasındaki İlişki. *Dokuz Eylül Üniversitesi İktisadi ve İdari Bilimler Fakültesi Dergisi, 27*(2), 25.
12. Lise, W., & Van Montfort, K. (2005). Energy consumption and GDP in Turkey: Is there a co-integration relationship?, In *EcoMod 2005 interantional conference on policy modeling*, İstanbul, Turkey.
13. TÜİK. Temel İstatistikler, Ulusal Hesaplar – Harcama Yöntemi ile GSYH – Gayrisafi yurtiçi hasıla, harcama yöntemiyle zincirlenmiş hacim, endeks ve değişim oranları, 1998 2017. http://www.tuik.gov.tr/UstMenu.do?mctod=temelist. Türkiye İstatistik Kurumu. Accessed 01 Feb 2019.
14. World Bank Open Data. *GDP (constant 2010 US$)*. https://data.worldbank.org/indicator/NY.GDP.MKTP.KD?end=2017&locations=TR&start=1972&view=chart. World Bank. Accessed 01 Feb 2019.
15. TCMB. Elektronik Veri Dağıtım Sistemi. Kurlar-Döviz kurları. Türkiye Cumhuriyeti Merkez Bankası. https://evds2.tcmb.gov.tr/index.php?/evds/serieMarket/#collapse_2. Türkiye Cumhuriyeti Merkez Bankası. Accessed 01 Feb 2019.
16. EIA. *Petroleum & other liquids – Data*. https://www.eia.gov/dnav/pet/hist/RWTCD.htm. U.S. Energy Information Administration. Accessed 01 Feb 2019.

Open Access This chapter is licensed under the terms of the Creative Commons Attribution 4.0 International License (http://creativecommons.org/licenses/by/4.0/), which permits use, sharing, adaptation, distribution and reproduction in any medium or format, as long as you give appropriate credit to the original author(s) and the source, provide a link to the Creative Commons license and indicate if changes were made.

The images or other third party material in this chapter are included in the chapter's Creative Commons license, unless indicated otherwise in a credit line to the material. If material is not included in the chapter's Creative Commons license and your intended use is not permitted by statutory regulation or exceeds the permitted use, you will need to obtain permission directly from the copyright holder.

Part III
Sustainable Organisations

Enhancing Social-Environmental-Economical Systemic Vision: Applying OLCA in a NGO

José Manuel Gil-Valle and Juan Pablo Chargoy-Amador

Abstract Emmaüs International a non-governmental organization (NGO) in the social and environmental sector had practiced, since its foundation – now more than 60 years – the recuperation of objects that others consider as waste. This activity had allowed collecting the funds to help the needy giving them the means to find their dignity that society had taken. Nowadays, the modes had changed, and these recovery activities had made of Emmaüs movement a well-known actor against the non-controlled waste "an environmental actor" working in the reuse and recycling. Given its environmental focus, Emmaüs has interest in assessing the environmental impacts of its own activities throughout the whole value chain. Therefore, an organizational life cycle assessment (O-LCA) study had been conducted as a test in one Emmaüs community. The study was realized in the framework of the road testing of the UNEP/SETAC Guidance on Organizational Life Cycle Assessment. It is important to mention that the avoided burdens assessment is not part of the O-LCA method.

1 Introduction

The Emmaüs community Etagnières, as a non-governmental organization (NGO) in the social and environmental sector, is interested in assessing the environmental impacts of its own activities throughout the whole value chain. Therefore, an organizational life cycle assessment (O-LCA) study was conducted. The study was performed in the framework of the road testing of the UNEP/SETAC Guidance on Organizational Life Cycle Assessment [1, 2].

Emmaüs' goals are of analytical, managerial and societal nature. The O-LCA study offer insights in internal operations as well as in other steps of the value chain, with a focus on wood board recycling. The results allow identifying environmental

J. M. Gil-Valle (✉)
LCI Member (Private advisor), Etagnières, Switzerland

J. P. Chargoy-Amador
Center for Life Cycle Assessment and Sustainable Design (CADIS), Life Cycle Management Director, Mexico City, Mexico

© The Author(s) 2022
Z. S. Klos et al. (eds.), *Towards a Sustainable Future - Life Cycle Management*,
https://doi.org/10.1007/978-3-030-77127-0_15

hotspots and set a reference for performance tracking over time. In a parallel study, the avoided burdens originated by the nature of the organization (recycling) will be analysed and compared with the results of the O-LCA. It is important to mention that the avoided burdens assessment is not part of O-LCA.

The study delivers the basis for environmental communications with stakeholders and reporting and allows showing environmental awareness with marketing purposes.

In general, the results of the study were analysed as an outcome of the road-testing phase of the Flagship initiative "LCA of Organizations" in the framework of the UNEP/SETAC Life Cycle Initiative and are publically available.

2 Materials and Methods

Using life cycle assessment (LCA) to quantify the environmental performance of products has become a global trend, since a comprehensive evaluation is achieved, considering all stages of the life cycle, as well as the different environmental problems, including the carbon footprint. The advantages and potential of LCA are not limited to a product application, and although the methodology was originally developed with this approach, its application at the organizational level is possible and is increasingly relevant.

The technical specification ISO/TS 14072:2014 Environmental management – Life cycle assessment – Requirements and guidelines for organizational life cycle assessment [3] describes the application of LCA with an organizational approach. In this way, it extends the application of ISO 14040 [4] and ISO 14044 [5] for all the activities of the organization, which means that the system evaluated covers the life cycle of the different products and operations within the same study.

O-LCA consists of the collection and evaluation of inputs, outputs and potential environmental impacts of the activities associated with an organization considered as a whole or portions of it, adopting a life cycle perspective.

ISO/TS 14072: 2014 provides details on:

- The application of LCA principles and methodology to organizations.
- The benefits that LCA can provide to the organization, using the methodology at the organizational level such as defining environmental aspects in the Environmental Management Systems ISO 14001: 2015, quantifying the environmental impact in an integral way and helping in strategic decision-making and prioritizing the actions that must be carried out to reduce the environmental impact of the organization.

O-LCA quantifies potential environmental impacts through a reporting flow, which is equivalent to the functional unit in a traditional LCA and is used as a reference. The system limits are defined by one of the following consolidation methodologies:

- Operational control
- Financial control
- Participation in shares (percentage of ownership)

In addition, O-LCA proposes two ways to perform data collection: the bottom-up approach and the top-down approach. In the first, the impact of the organization will be calculated with the sum of the LCA of each of the products it manufactures. This implies a collection of data broken down by product, which can be extremely complex for organizations with large portfolios. In the case of the top-down approach, the inputs and outputs of the system can be collected as a whole, by production plant (site) or even by business group. This approach eases the collection of information and allows disaggregation of the results according to the information needs of the organization.

O-LCA can be used as an input for environmental communication, especially for monitoring the environmental performance of the organization over time (performance tracking).

3 Results

3.1 Goal and Scope

The assessed organization was a local Emmaüs community, located in Etagnières, Switzerland, during 1 year from January 2015 to December 2015. The reporting flow was the annual sales expressed in mass (kg).

The system boundary considered a cradle-to-gate approach for the inputs and outputs necessary for each of the activities included, extended by considering the transport of sold goods by the costumers. The production and first use of products are not considered, as well as the use and end of life of the sold recycled materials. The activities considered are categorized into indirect upstream activities and direct activities. Supporting activities like the organization's buildings and employee commuting were considered. System boundary is depicted in Fig. 1.

3.2 Inventory Analysis

A top-down screening approach was used as first approximation to obtain a basis for future studies. Transport data is collected with higher granularity and disaggregated into trucks transport, direct donor transport and customer transport.

Energy data was disaggregated in energy production on site and electrical – solar.

Both generic and specific data were used. The source is on-site, from literature, statistics and databases. A data quality scheme was used with the following criteria:

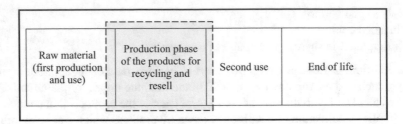

Fig. 1 Emmaüs community system boundary

Table 1 Impact assessment results

Damage (Pt)	Direct activities	Indirect activities	Total
Climate change, human health	2294	11,594	13,888
Ozone depletion	0,000	0,004	0,004
Human toxicity	0,073	1513	1586
Photochemical oxidant formation	0,000	0,000	0,000
Particulate matter formation	0,543	1704	2247
Ionizing radiation	0,000	0,145	0,145
Climate change, ecosystems	1451	7336	8788
Terrestrial acidification	0,004	0,009	0,014
Freshwater eutrophication	0,000	0,008	0,008
Terrestrial ecotoxicity	0,003	0,021	0,025
Freshwater ecotoxicity	0,000	0,042	0,042
Marine ecotoxicity	0,000	0,008	0,008
Agricultural land occupation	0,000	0,493	0,494
Urban land occupation	0,000	0,095	0,095
Natural land transformation	0,000	0,409	0,409
Metal depletion	0,003	0,950	0,953
Fossil depletion	0,003	11,615	11,618
Total	4378	35,953	40,331

reliability, completeness, temporal correlations, geographical correlation and further technological correlation.

3.3 Impact Assessment

The impact assessment method ReCiPe Endpoint (H) [6] was applied. The main impacts have been detected in the categories climate change, human health and ecosystem followed by fossil depletion and particle matter formation. The impacts related to the transportation of sold materials represent an overall contribution of 41%. Impact assessment results are depicted in Table 1.

4   Discussion

The assessment with the ReCiPe Endpoint method allowed identifying environmental hotspots in the impact categories climate change, human health and ecosystem, followed by fossil depletion and particle matter formation. Electricity production, organization's buildings and transport of purchased goods are found being relevant activities. Actions to reduce transport-related impacts, such as selling points next to potential customers and online sales, are recommended.

The main limitations of the study consist in the exclusion of certain capital goods such as trucks and the boiler. The same applies for cleaning products, medicines, gardening products and personal care products that could be analysed in the future because of the potential effects of micropollutants. Facilities as kitchen, green and gardening areas were not included since they were already targeted in the framework of our food recuperation programme. Moreover, the use and end-of-life phase of the sold recycled products are not considered in this study.

Through O-LCA study, the hotspots could be detected. This could help improving the image of the community as a main actor regarding environmental activities. Emmaüs' study was a pilot and serves as example for other Emmaüs communities around the world. As first application in an NGO, Emmaüs' O-LCA experience has the great potential of being a landmark for environmental assessment activities among charitable organization.

5   Conclusions

O-LCA is useful in detection of the main environmental impact categories and their contribution concerning indirect and direct activities. A performance tracking of the mentioned activities could be established from this study on.

The study delivers the basis for the communication of "Sustainable Development Issues" with stakeholders (customer, services providers and partners) and reporting.

A basic model to apply the O-LCA methodology had been established in an Emmaüs recycling community that could be applied in other Emmaüs communities in the future.

The tools developed to apply this methodology were designed with the aim of supporting recycling communities around the world and the whole Emmaüs organization to evaluate and to reduce their environmental impacts in their own communities but also in the regions where they operate, thus positively affecting local development.

Further applications of the study are being considered. First, the data collected could be used in the future as environmental data basis for a formal Environmental Management System (EMS). Second, the Emmaüs community could serve as a pilot project as O-LCA is concerned. In fact, further recycling communities worldwide could apply the methodology in the future, thus enabling an assessment of the whole organization or a broader part of it.

From this perspective, Emmaüs is a first mover in the NGO sector.

References

1. UN environment. (2017). *Road testing organizational life cycle assessment around the world.* Life Cycle Initiative.
2. Guide on Organizational Life Cycle Assessment (2015, English, 148 pages).
3. ISO/TS 14072:2014 Environmental management – Life cycle assessment – Requirements and guidelines for organizational life cycle assessment.
4. ISO 14040:2006 Environmental management – Life cycle assessment – Principles and framework.
5. ISO 14044:2006 Environmental management – Life cycle assessment – Requirements and guidelines.
6. Huijbregts, M., Steinmann, Z., Elshout, P., Stam, G., Verones, F., Vieira, M., Zijp, M., Hollander, A., & van Zelm, R. (2017). ReCiPe2016: A harmonised life cycle impact assessment method at midpoint and endpoint level. *The International Journal of Life Cycle Assessment, 22*(2), 138–147. https://doi.org/10.1007/s11367-016-1246-y

Open Access This chapter is licensed under the terms of the Creative Commons Attribution 4.0 International License (http://creativecommons.org/licenses/by/4.0/), which permits use, sharing, adaptation, distribution and reproduction in any medium or format, as long as you give appropriate credit to the original author(s) and the source, provide a link to the Creative Commons license and indicate if changes were made.

The images or other third party material in this chapter are included in the chapter's Creative Commons license, unless indicated otherwise in a credit line to the material. If material is not included in the chapter's Creative Commons license and your intended use is not permitted by statutory regulation or exceeds the permitted use, you will need to obtain permission directly from the copyright holder.

LCA in the Field of Safety at Work: A New Engineering Study Subject

Boris Agarski, Dejan Ubavin, Djordje Vukelic, Milana Ilic Micunovic, and Igor Budak

Abstract Life cycle assessment (LCA) is a standardised and comprehensive approach for evaluation of environmental impacts within the material and energy flows associated with various human activities and through the life cycle stages. Besides environmental impact evaluation, with LCA, costs, social impacts, impacts on workers, organisations and others can also be assessed. This paper focuses on development of educational framework for evaluation of occupational safety based on LCA. The goal is to develop a new study subject "LCA in the field of safety at work" for the occupational safety engineering master study programme at the Faculty of Technical Sciences in Novi Sad. New study subject is based on LCA approaches that evaluate the occupational safety and impact on workers. Based on the previous research of LCA in the field of occupational safety, the goal, outcome, content and realisation are defined for the new study subject.

1 Introduction

Life cycle assessment (LCA) has been in education process at the University of Novi Sad for more than 20 years, since the foundation of the Department of Environmental Engineering at the Faculty of Technical Sciences. The starting point was a teaching topic within the environmental engineering study programme, the subject mechanical engineering in environmental protection. Today, LCA is studied in several courses at bachelor, master and PhD levels of environmental, occupational safety, mechanical and civil engineering study programmes. The result is a growing number of bachelor, master and PhD theses in the field of LCA, eco-labelling and eco-design. Considering the importance of occupational safety in engineering and aiming to fulfil the expectations of organisations operating on the labour market, besides the environmental engineering, since 2010 occupational safety engineering study programme has been established at the Faculty of Technical Sciences.

B. Agarski (✉) · D. Ubavin · D. Vukelic · M. I. Micunovic · I. Budak
Faculty of Technical Sciences, University of Novi Sad, Novi Sad, Serbia
e-mail: agarski@uns.ac.rs

© The Author(s) 2022
Z. S. Klos et al. (eds.), *Towards a Sustainable Future - Life Cycle Management*,
https://doi.org/10.1007/978-3-030-77127-0_16

173

Besides environmental LCA, life cycle costing and social LCA (S-LCA) emerge in order to provide sustainable LCA, where S-LCA is the youngest methodology. Within the S-LCA [1], impact on workers' health and safety during the life cycle is a group of stakeholder impact categories that can provide information on accident rates at workplace (non-fatal and fatal), occurrence of various diseases and injuries, disability-adjusted life years (DALYs), presence of safety measures, etc. Working environment LCA (WE-LCA) [2] aim to compile and evaluate potential working environmental impacts on humans of a product system throughout its life cycle. The impact categories in WE-LCA can be expressed through evaluation of potential accidents and diseases: fatal accidents, total number of accidents, central nervous system function disorder, hearing damages, cancer, musculoskeletal disorders, airway diseases (allergic and non-allergic), skin diseases and psychosocial diseases. Furthermore, damage to human health attributable to the work environment can be assessed as DALYs [3].

Table 1 provides several approaches for WE-LCA. Schmidt et al. [2] developed one of the first WE-LCA approaches. This WE-LCA approach is based on EDIP life cycle impact assessment method and contains a small life cycle inventory (LCI) database with more than 80 activities. Pettersen and Hertwich [4] focused on evaluation of safety issues related to offshore crane lifts working environment. Kim and Hur [5] developed two working environment indicators in context of LCA: occupational health and occupational safety. One of the first S-LCA case studies that followed the UNEP/SETAC S-LCA guidelines [1] was presented transparently and in detail was realised by Ciroth and Franze [6]. Group of authors [3, 7] provided two papers published in 2013 and 2014 and used national occupational safety and health industry statistics for United States of America to express the impact on working environment through the WE-DALY units. For WE-DALY indicator, they [3] provided 127 working environment characterisation factors linked with various industry sectors. Kijko et al. [8] also used DALY units to assess health impacts from occupational exposure to chemicals. Khakzad et al. [9] used LCA and quantitative risk assessment methods in parallel to obtain the environmental and safety assessment. Monetary valuation, Canadian dollar (CAD) units were used for both methods in order to have comparable outputs from LCA and quantitative risk assessment.

It can be noted that all approaches in Table 1 have the following common characteristics:

- Compatible with ISO 14040 LCA phases and environmental LCA.
- National statistic records of safety issues through the industrial sectors are used to evaluate safety at work, or to assess the risk of injuries and illness.
- Although developed on national level, all approaches have the potential for universal worldwide use.

Considering that the working environment indicators are relatively new topic in LCA, and that research in the field of S-LCA is an actual topic nowadays, this paper focuses on development of educational framework for LCA in the field of safety at work and working environment in LCA. The goal of this paper is to develop a new study subject on a master study programme of occupational safety engineering at

Table 1 LCA approaches to evaluate safety at work

Approach	Working environment in life cycle assessment	Human health impact indicator for offshore crane lifts	Hybrid input-output analysis
Acronym	WE-LCA	–	Hybrid IOA
Reference	[2]	[4]	[5]
Developing basis	EDIP[b] method	LCA and DALY[d] units	LCA and IOA method
Problem-solving	Impacts on workers/universal	Development of a human health impact indicator for offshore crane lifts	Assessment of occupational health and safety
Geography	Denmark	United Kingdom	Korea
Characterisation	Based on statistics on work-related accidents and reported diseases from the Danish Labour Inspectorate and Statistics on the amounts of produced goods in Denmark	Based on number of crane lift incident injuries and expressed in DALY per crane lift	Linking the LCI[a] data with 28 basic industrial sectors classified by the Bank of Korea for occupational health and Korea Occupational Safety and Health Agency for occupational safety
No. of impact categories	10 – fatal accidents, total number of accidents, hearing damages, cancer, musculoskeletal disorders, airway diseases (allergic), airway diseases (non-allergic), skin diseases, psychosocial diseases, CNS function disorder	1 – health burden per crane lift	2 – occupational health (number of workers affected by certain hazardous items) and occupational safety (number of workers at certain magnitude of disability)
Normalisation	Yes – 2 sets: Danish population (person equivalents) and Danish work force (worker equivalents)	Yes – number of lifts performed per hour	Yes – total national lost work days from the occupational diseases by hazardous items during the given period of time divided by the total number of the workers
Developed and provided LCI[a] database	Yes – more than 80 activities based on DB93[c] industry sectors	No	No

[a]LCI, life cycle inventory
[b]EDIP, Danish Environmental Agency
[c]DB93, Danish nomenclature for industry sectors (identical to the EU NACE-code system)
[d]*DALY* disability-adjusted life years

the Faculty of Technical Sciences in Novi Sad in order to produce occupational safety engineers that will be able to assess the impacts on workers' health and safety with LCA approach.

2 Methodology

The study programme of the graduate master academic studies in Occupational Safety Engineering presents the continuation of the undergraduate academic studies of Occupational Safety Engineering at the Faculty of Technical Sciences, University of Novi Sad [10]. Engineering and technical disciplines are incorporated into the realisation of the curriculum of the undergraduate and graduate academic studies of Occupational Safety Engineering, thus representing a highly multidisciplinary and interdisciplinary programme. The study programme prerequisites for the enrolment are completed undergraduate studies with at least 240 ECTS and the passed enrolment examination. General information on Master in Occupational Safety Engineering study programme are provided in Tables 2 and 3.

Distribution of ECTS points in master academic studies in occupational safety engineering is provided in Fig. 1. The other study subjects (curriculum) on occupational safety engineering study programme tackle topics such as hazardous materials and hazardous waste, occupational risk assessment, statistical advanced models, occupational medicine, chemical risk assessment of fire and explosion, system regulations and EU practice in occupational health and safety, occupational noise and human vibration in industry, accidental risk management and the environment, product safety and user/consumer protection and sociological and legal aspects of occupational safety. On the other side, none of the current subjects cover the safety at work from life cycle perspective.

According to the previously defined study subject topic, the goal, outcome, content and realisation of new study subject will be defined in results section.

3 Results

Based on the previous literature, the new study subject LCA in the field of safety at work has to cover the following topics (Fig. 2):

- LCA according to ISO 14040 and 14044 international standards
- Relationship between WE-LCA and other LCA approaches: the environmental LCA, S-LCA, life cycle costing organisational LCA and sustainability LCA
- S-LCA for workers stakeholder group: goal and scope definition, S-LCI, social life cycle impact assessment methods and interpretation
- Software support for S-LCA: S-LCA software and S-LCI databases
- Evaluation of products life cycle impact on workers through WE-DALY approach

Table 2 LCA approaches to assess safety at work (continued)

Approach	Social life cycle assessment	Work environment disability adjusted life year	Occupational LCA	Accident risk-based life cycle assessment
Acronym	S-LCA	WE-DALY	–	ARBLCA
Reference	[6]	[3, 7]	[8]	[9]
Developing basis	LCA	LCA and DALY units	LCA and DALY	LCA and quantitative risk assessment
Problem-solving	Evaluation of social impacts through the product's life cycle	Waste management – landfilling and incineration	Assessment of health impacts from occupational exposure to chemicals	Green and safe fossil fuel selection
Geography	Worldwide	United States of America	North American	Canada/ potentially worldwide
Characterisation	Assessment of the performance of the sectors and companies, respectively, based on the status of the indicators taking the performance of the sector/company in relation to the situation in the country/region into account	Characterisation factors are obtained from US industry-level occupational safety and health data (work-related fatal and non-fatal injuries and illnesses) and the physical quantities of goods produced by these industries	Based on labour hours and indoor intake concentration	IPCC[d]
No. of impact categories/ indicators	8 – within workers' stakeholder category, the subcategories are the following: freedom of association and collective bargaining, child labour, forced labour, fair salary, working time, discrimination, health and safety, social benefits/ social security	1 – work environment DALY[b] (WE-DALY)	1 – occupational exposure to chemicals expressed in DALY/h	2 – GHG[e] (CO_2) emissions converted to CAD[f] by carbon tax for LCA, and 5 risk loss categories in CAD[f]

(continued)

Table 2 (continued)

Approach	Social life cycle assessment	Work environment disability adjusted life year	Occupational LCA	Accident risk-based life cycle assessment
Normalisation	Yes – each subcategory is assessed twice with a colour system ranging from very good performance to very poor performance and very negative impacts to positive impacts	No	No	British Colombia province carbon tax (30 CAD[f] per metric ton of CO_2 equivalent)
Developed and provided LCI[a] database	LCI database is not provided in the particular study; however, S-LCA databases exist	Yes – 127 WE characterisation factors linked with NAICS[c] industry sectors	Yes – for various NAICS [c] industry sectors, characterisation factors have been developed for 19069 organic chemical/sector combinations	None

[a]LCI, life cycle inventory
[b]*DALY* disability-adjusted life years
[c]*NAIC*, North American Industry Classification System
[d]*IPCC* Intergovernmental Panel on Climate Change
[e]*GHG* greenhouse gases
[f]*CAD* Canadian dollar

Table 3 General information on master in occupational safety engineering study programme [10]

Type of studies	Master academic studies
Academic degree	Master in Occupational Safety Engineering (M. Occ.Saf.Eng.)
Educational field	Technical-Technological Science
Scientific, professional or art field	Environmental and Occupational Safety Engineering
Duration (year/sem)	1 year/2 semesters
Total European Credit Transfer System (ECTS) points	60
Web address containing study programme information	http://www.ftn.uns.ac.rs

Fig. 1 Distribution of ECTS points in master academic studies in occupational safety engineering

Fig. 2 Topics in study subject LCA in the field of occupational safety

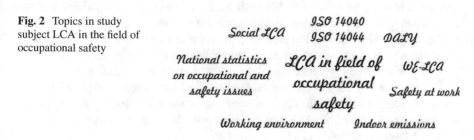

- Evaluation of products life cycle impact on workers through the WE-LCA approach

Fundamentals for teaching will certainly include recommendations for LCA from ISO 14040 and 14044. These standards provide basics for environmental LCA and are nowadays incorporated in other LCA approaches. Historical development, similarities and differences between the various LCA approaches are interesting starting point for better understanding of LCA in the field of safety at work. Within S-LCA, besides other social issues, evaluation of occupational safety is expressed through the workers stakeholder impact category. Software support for S-LCA enables practical calculations of social impacts, supply chain modelling and connection between the industry sectors and countries. Therefore, S-LCA software can be used for performing exercises in computer classrooms with students. WE-DALY and WE-LCA approaches have their LCI database which also can be used for exercises in computer classrooms with students.

The new subject LCA in the field of safety at work on a master study programme of occupational safety engineering at the Faculty of Technical Sciences in Novi Sad has been developed and applied for the accreditation programme for the new 2020/2021 academic year. Goal, outcome, content and realisation of this subject are provided in the following part:

- Goal: Acquisition of knowledge, competences and academic skills in field of safety at work and product's life cycle. Development of creative capabilities, academic and practical skills for implementation of life cycle assessment of processes and products from aspect of impact on the worker;

- Outcome: Ability to solve real problems in the field of life cycle assessment of product's impact on worker. Mastering methods and procedures for life cycle assessment of product's impact on worker. Development of skills for life cycle assessment of product's impact on worker with respecting the sustainable development principles. Ability to critically and self-critically think within interpretation of product's and process's life cycle assessment results.
- Content: Product's life cycle. Life cycle assessment in the field of environmental protection and safety at work. Sustainable development, economic, social and environmental dimension within the life cycle assessment. Defining goal and scope of study. Life cycle inventory. Life cycle inventory databases. Life cycle impact assessment on worker. Methods for life cycle impact assessment of products and processes on worker. Interpretation of results.
- Realisation: Lectures are interactive in the form of lectures, auditory, laboratory and computer practice. During the lectures, theoretical part of the course is presented followed by typical examples for better understanding. During the auditory practice, typical problems are solved and the knowledge is deepened. During the computer practice, information communication technologies are applied in order to master the knowledge of the observed field. Besides lectures and practice, consultations are held on a regular basis.

Besides the lectures, this study subject is based on exercises where students can obtain practical knowledge. The exercises have to be based on interactive relationship between the lecturer and students and use of modern educational equipment, computers and the Internet. Mastering methods from this study subject will enable students to perform and develop skills for LCA of product's and process's impact on worker health and safety.

4 Conclusions

Although the environmental LCA is well known, the social LCA and LCA in the field of safety at work are starting to gain their momentum in scientific community. The new study subject LCA in the field of safety at work on a master study programme of occupational safety engineering at the Faculty of Technical Sciences in Novi Sad aims to enable students to master these methods and to perform and develop skills for LCA of product's and process's impact on worker health and safety. The objective is to achieve student's scientific competencies and academic skills in the field of LCA and occupational safety. One of the specific objectives is to develop students' awareness of the need for continuous education in the field of occupational safety and the development of a society in general.

The educational framework in this paper is developed for the purposes of occupational safety engineering study programme at the Faculty of Technical Sciences in Novi Sad. However, this framework can be applied at other study programmes and universities with certain modifications according to their specific needs. Further development directions will be detected after implementation of LCA in the field of safety at work study subject.

References

1. UNEP/SETAC, Guidance for Social Life Cycle Assessment of Products. Life-Cycle Initiative, United Nations Environment Programme and Society for Environmental Toxicology and Chemistry, Paris, France, 2009.
2. Schmidt, A., Poulsen, P. B., Andreasen, J., Fløe, T., & Poulsen, K. E.(2004). *LCA and the working environment.* Environmental Project No. 907, Danish Environmental Protection Agency.
3. Scanlon, K. A., Lloyd, S. M., Gray, G. M., Francis, R. A., & LaPuma, P. (2014). An approach to integrating occupational safety and health into life cycle assessment: Development and application of work environment characterization factors. *Journal of Industrial Ecology, 19*(1).
4. Pettersen, J., & Hertwich, E. G. (2008). Occupational health impacts: Offshore crane lifts in life cycle assessment. *International Journal of Life Cycle Assessment, 13,* 440–449.
5. Kim, I., & Hur, T. (2009). Integration of working environment into life cycle assessment framework. *International Journal of Life Cycle Assessment, 14,* 290–301.
6. Ciroth, A., & Franze, J., LCA of an ecolabeled notebook – Consideration of social and environmental impacts along the entire life cycle, Berlin 2011.
7. Scanlon, K. A., Gray, G. M., Francis, R. A., Lloyd, S. M., & LaPuma, P. (2013). The work environment disability-adjusted life year for use with cycle assessment: A methodological approach. *Environmental Health, 12*(21).
8. Kijko, G., Margni, M., Parovi-Nia, V., Doudrich, G., & Jolliet, O. (2015). Impact of occupational exposure to chemicals in life cycle assessment: A Novel characterization model based on measured concentrations and labor hours. *Environmental Science & Technology, 49,* 8741 8750.
9. Khakzad, S., Khan, F., Abbassi, R., & Khakzad, N. (2017). Accident risk-based life cycle assessment methodology for green and safe fuel selection. *Process Safety and Environmental Protection, 109,* 268–287.
10. FTS – Master in Occupational Safety Engineering study programme. http://www.ftn.uns.ac.rs/n1473594640/safety-at-work (Accessed 04.02.2020)

Open Access This chapter is licensed under the terms of the Creative Commons Attribution 4.0 International License (http://creativecommons.org/licenses/by/4.0/), which permits use, sharing, adaptation, distribution and reproduction in any medium or format, as long as you give appropriate credit to the original author(s) and the source, provide a link to the Creative Commons license and indicate if changes were made.

The images or other third party material in this chapter are included in the chapter's Creative Commons license, unless indicated otherwise in a credit line to the material. If material is not included in the chapter's Creative Commons license and your intended use is not permitted by statutory regulation or exceeds the permitted use, you will need to obtain permission directly from the copyright holder.

Setting Internal Price of Environmental Criteria, the Good Way to Transform Organization?

Stéphane Morel, Nabila Iken, and Franck Aggeri

Abstract In this communication, we present some lessons learned on the construction of an internal carbon price by businesses, based on the four-dimensional framework of the Carbon Disclosure Project. We illustrate the scheme with the example of a car manufacturer. Based on grey literature and the conclusions of exchanges with various companies, we discuss the different dimensions of the CDP framework within the scope of the automotive sector. We also analyse the various risk and success factors associated with the carbon pricing tool at organizational, tooling, business and cultural levels within a car manufacturer. We conclude that the carbon pricing tool requires many design choices and a reflection on the company's objective regarding climate change mitigation.

1 Introduction

Whether in the form of taxes, emissions trading systems or other mechanisms, there are currently 57 carbon pricing initiatives implemented or scheduled for implementation worldwide, covering 46 national jurisdictions [1]. However, the carbon prices emanating from them are very disparate and often not commensurate with the issues at stake. Indeed, they vary from less than US1\$/tCO2e (Poland carbon tax) to 127US\$/tCO2e (Sweden carbon tax), with 51% of emissions priced less than 10 US\$/tCO2e. Therefore, some companies are proactively adopting non-regulatory (so-called internal) carbon prices. Even though this practice involved more than 1300 companies in 2017 [2], there is little research on how this price is constructed in practice and deployed internally by companies, which will be the subject of this communication.

S. Morel (✉)
Alliance Technology Development, Renault sas, Guyancourt, France
e-mail: stephane.s.morel@renault.com

N. Iken · F. Aggeri
CGS-i3 Mines-ParisTech, PSL University, Paris, France

© The Author(s) 2022
Z. S. Klos et al. (eds.), *Towards a Sustainable Future - Life Cycle Management*,
https://doi.org/10.1007/978-3-030-77127-0_17

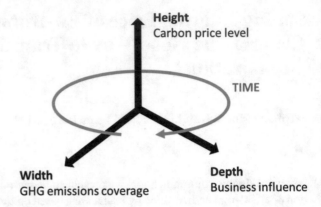

Fig. 1 Four dimensions of internal carbon pricing

2 Method

In order to grasp the implications of setting an internal carbon price in a company, we based our analysis on the Carbone Disclosure Project four-dimensional framework [3], illustrated in Fig. 1. Indeed, we considered that the integration of an ICP[1] by a company is determined by its (i) height, the carbon price level adopted; (ii) width, the emissions coverage in terms of indirect and/or direct greenhouse gas emissions and company's activities concerned; (iii) time, evolution of carbon pricing strategy through time; and (iv) depth. the business influence (informative or decisional ICP? In which form?). In the following, we describe our findings on each of the dimensions described above, in the case of a car manufacturer.

3 Results

Through our study of the grey literature as well as corporate practices in the private sector, our objective is to strengthen managerial knowledge about carbon pricing. Our results therefore make it possible to move a little further towards putting carbon pricing into practice, by highlighting various avenues for reflection in the case of a car manufacturer.

3.1 Height: Carbon Price Level

To reflect the cost of greenhouse gas emission-related externalities in the economic system, the monetary valuation of carbon has been the subject of concern among economists, public authorities and scientists [4]. This has given rise to a multitude of possible forms and values of carbon, which is reflected in the current regulatory

---

[1] Internal carbon price.

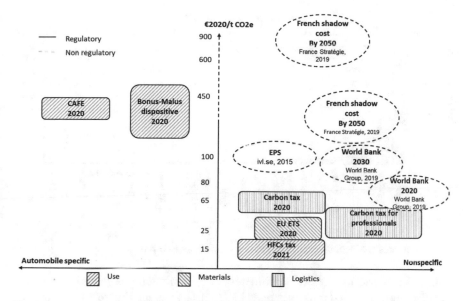

Fig. 2 Regulatory carbon pricing initiatives affecting carmakers in France and other carbon prices

landscape. Because the automotive sector is one of the largest sources of greenhouse gas emissions in Europe (72% of transport CO_2 emissions [5]), some regulatory measures directly target this industry. Figure 2 presents the French regulatory context, where the dates in bold represent the date of application of the measure for regulatory prices, or the time horizon within which the prices should be applied (for non-regulatory prices).

The choice of the carbon price therefore comes down to a positioning in relation to regulations (a degree of anticipation), but also to the company's ambition in terms of the objectives to be achieved (alignment with best market practice, alignment with the 2 °C objective or another company-specific objective). There is also a whole dimension related to internal feasibility, depending on whether the carbon pricing initiative comes from top management, in which case it is a question of deployment, or elsewhere in the company, where it is more of a negotiation process with the decision-makers.

3.2 Width: Emissions Coverage

Figure 2 shows that the carbon pricing regulatory initiatives tend to reduce CO_2 emissions on the use phase of vehicles, much less the emissions in the upstream stages of the vehicle's life cycle. This can lead to a transfer of pollution to phases of the life cycle that are not covered by these regulations, in particular materials production. For this reason, the use of an internal carbon price makes sense within the scope of materials, whether to anticipate regulatory changes or to prevent the transfer of pollution.

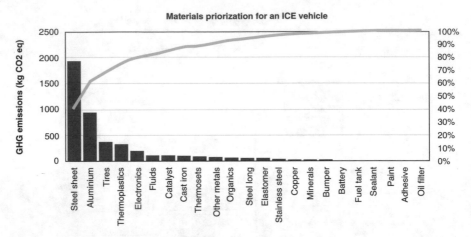

Fig. 3 Materials production greenhouse gas emissions in an ICE vehicle

In order to prioritize relevant perimeters of carbon pricing of materials for a car manufacturer, we based ourselves on vehicles' LCA results. Figure 3 illustrates the greenhouse gas emissions due to the production of different materials for an electric vehicle, without the Li-ion battery. Figure 3 shows the same for an ICE vehicle.

On this basis, we have selected the following priority perimeters.

Besides, LCAs are conducted with a scope 3 perimeter, which means that both direct and indirect emissions are considered through the whole life span of the vehicles.

3.3 Depth: Business Influence

For the carbon price to play the role of a transformative tool, it must be embedded in the company's decision-making processes. This raises the question of making it consistent with existing tools and calls for examples of possible use.

For this reason, we conducted a survey with 13 companies that disclose their use of management tools involving monetary valuation of environmental externalities (including carbon). This allowed us to identify the following four categories of tools:

3.3.1 Assessing the Environmental P&L[2]

In the natural capital valuation movement pioneered by PUMA [6], several companies have calculated and communicated their Environmental P&L or Integrated P&L (including social externalities). It is a company's monetary valuation of its

---

[2] Profit and loss.

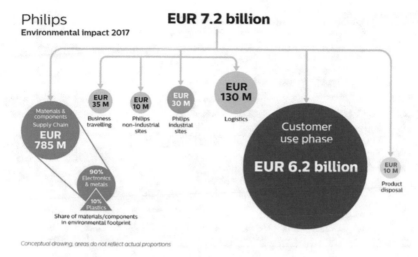

Fig. 4 Philips' 2017 Environmental Profit and Loss accounting

environmental impacts to see their magnitude, disclose them to stakeholders and possibly guide the company's strategy. Figure 4 shows the result of the EP&L calculation made by Philips in 2017 [7].

3.3.2 Including the Cost of Externalities in the TCO[3]

One way to consider the price of carbon in business decisions is to integrate it into cost indicators, such as TCO. Volvo Bus company applied this method to compare between electric and diesel buses in Sweden (Fig. 5), by including environmental and social externalities in the TCO calculation [8].

3.3.3 Including a Shadow Price in the NPV

Another method identified is the integration of a shadow price in the calculation of indicators for investment choices such as the net present value (NPV). This is a way of applying a pricing scenario on a resource or pollutant (in this case carbon). For example, Dow Chemical used this approach to introduce the hidden cost of water into their infrastructure investment choices [9].

---

[3] Total cost of ownership.

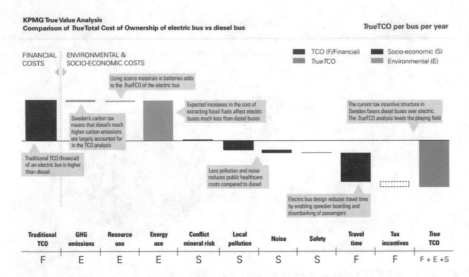

Fig. 5 Volvo Bus' true TCO of electric buses compared with diesel buses

3.3.4 Integrating the External Costs in the Portfolio Strategy

It is also possible to introduce external costs in general – or carbon price in particular – into business strategy through a portfolio management tool, in order to gradually eliminate the most impactful products from the portfolio and replace them with the most virtuous ones. Such a product-oriented approach has been developed by the chemical company Solvay [10] under the name of Sustainable Portfolio Management. Figure 6 shows how the SPM allows mapping the different PACs[4] in the portfolio in two dimensions: (i) market alignment, which is a qualitative estimation of market early signals related to sustainability in the chemical industry, and (ii) operations vulnerability, which is the ratio of the external cost related to the product and its sales value. The blue colour scale represents the turnover associated with the PAC.

Based on the available materials in grey literature and our discussions with the companies, we classified the previous tools typical use according to these two axes:

- **External *versus* Internal**: indeed, some tools are rather designed for communication purposes with external stakeholders and are often mobilized as a means of enriching the sustainability report. On the contrary, some tools are rather intended to guide corporate strategy, investment or portfolio choices. However, it doesn't prevent a tool from playing both roles at the same time.
- **Prospective *versus* Retrospective**: if the tools use data from past activities, they are retrospective and therefore allow an a posteriori evaluation of the company's

---

[4] Product in an application.

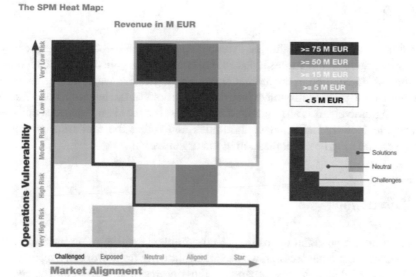

Fig. 6 Solvay's Sustainable Portfolio Management

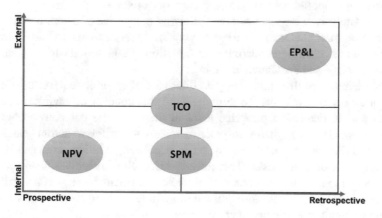

Fig. 7 Typical tools' use by businesses

activities. Similarly, if they are based on future projections (e.g. cost forecasts or future technological developments), then they are prospective.

Figure 7 shows the position of each tool described according to their typical use by the companies.

3.4   Time

The time dimension highlights the dynamic nature of the carbon pricing process in a company. Indeed, this makes it possible to envisage the construction of a roadmap for the implementation of an internal price in a progressive way, starting, for example, with a low price to minimize internal oppositions at the beginning and increasing it progressively. It is also more realistic to test the tool in a reduced scope (pilot project) to identify risks and opportunities and refine the tool's design choices before considering its generalization in the organization.

4   Discussion

To illustrate the potential oppositions to the implementation of an internal carbon price within a car manufacturer, we used the following framework as a reading grid. We considered that an induced change in the routines of an actor – or a category of actors – can be subdivided into changes in its (i) culture, (ii) competences, (iii) organization and (iv) tools. This allows identifying the possible oppositions and adapting the proposed solutions to each category of actors.

In our analysis, we considered the following categories of actors based on their influence on materials use: materials buyers, materials experts and environmental experts. Our conclusions concern the introduction of an internal carbon price in the form of an NPV and are shown in Table 1.

Table 1 indicates that introducing an internal carbon price requires the development of an often new expertise to understand this concept of environmental economics and to determine the price level in line with the company's objectives. However, moving from theory to practice means for different actors accepting to change the time horizon of decisions, by incorporating a hidden cost that is a kind of anticipation of future risks. This may conflict with immediate financial objectives, hence the need to reflect on both the relevant perimeter (e.g. considering that R&D and innovation gives more latitude to include the long term in decisions) and also the discourse and rhetoric that accompanies this tool.

Table 1 Analysis of the change due to the integration of an internal carbon price in the form of a NPV

Materials buyers		Familiar with the NPV tool NPV is already a decision criterion Difficulty to consider a shadow cost on the same level with internal costs (cultural gap)
Environmental experts		Already aware of environmental issues Familiar with environmental impact assessment tools Need for learning in the field of carbon pricing
Materials experts		Are used to favouring materials with the best technical-economic performance Need to be more in touch with environmental experts

5 Conclusion

In this communication, we exposed some learnings about the practice of internal carbon pricing, and its potential application in the automotive sector. We showed that the choice of the height of the price was ultimately a choice of target concerning the reduction of CO_2 emissions over a given time horizon. We also demonstrated that the perimeter of materials was a relevant field of application for a car manufacturer and proposed different forms of integration based on companies' practices. However, we have also illustrated the potential difficulties in implementing this tool in a company, especially if it is not a top management initiative. This is why this tool must be an element of a more global approach involving the dissemination of long-term strategic thinking with regard to sustainability issues.

References

1. The World Bank. (2019). *State and trends of carbon pricing*.
2. CDP. (2017). *CDP Carbon Majors Report 2017*, p. 16.
3. CDP. HOW-TO GUIDE TO CORPORATE INTERNAL CARBON PRICING: Four Dimensions to Best Practice Approaches. Generation Foundation, ECOFYS, CDP, Sep. 2017.
4. Tol, R. S. J.. (2008). *The Social cost of carbon: Trends, outliers and catastrophes*, p. 24.
5. European Environment Agency. (2019). *Greenhouse gas emissions from transport in Europe*.
6. PUMA. (2011). *Annual and sustainability report*.
7. Philips Innovation Services. (2018). *Growing trend in environmental profit & loss accounting: How to reap the benefits*. Philips Innovation Services.
8. Volvo and KPMG. (2015). *True value case study*, Volvo Group.
9. Shipp, E. (2017). *Natural capital protocol: Case study for dow chemical*, p. 2.
10. Solvay. (2017). *Sustainable portfolio management guide: Driving long-term sustainable growth*.

Open Access   This chapter is licensed under the terms of the Creative Commons Attribution 4.0 International License (http://creativecommons.org/licenses/by/4.0/), which permits use, sharing, adaptation, distribution and reproduction in any medium or format, as long as you give appropriate credit to the original author(s) and the source, provide a link to the Creative Commons license and indicate if changes were made.

The images or other third party material in this chapter are included in the chapter's Creative Commons license, unless indicated otherwise in a credit line to the material. If material is not included in the chapter's Creative Commons license and your intended use is not permitted by statutory regulation or exceeds the permitted use, you will need to obtain permission directly from the copyright holder.

Part IV
Sustainable Markets and Policy

Metal and Plastic Recycling Flows in a Circular Value Chain

Sasha Shahbazi, Patricia van Loon, Martin Kurdve, and Mats Johansson

Abstract Material efficiency in manufacturing is an enabler of circular economy and captures value in industry through decreasing the amount of material used to produce one unit of output, generating less waste per output and improving waste segregation and management. However, material types and fractions play an important role in successfulness of recycling initiatives. This study investigates two main fractions in automotive industry, namely, metal and plastic. For both material flows, information availability and standards and regulations are pivotal to increase segregation, optimize the collection and obtain the highest possible circulation rates with high quality of recyclables. This paper presents and compares the current information flows and standards and regulations of metals and plastics in the automotive value chain.

1 Introduction

In today's value chain, where production rate and correlated resource and energy consumption constantly increase, efficient and effective use of resources is imperative. In addition, recent concerns regarding non-renewable resources and environmental burden of extracting and producing products from virgin raw materials have been published in several reports and scientific publications such as [1–4]. Material efficiency is an approach within circular economy and resource efficiency to regain

S. Shahbazi (✉)
RISE IVF – Research Institutes of Sweden, Stockholm, Sweden
e-mail: sasha.shahbazi@ri.se

P. van Loon
Chalmers Industriteknik, Göteborg, Sweden

M. Kurdve
RISE IVF – Research Institutes of Sweden, Stockholm, Sweden

Department of Technology, Management and Economics, Chalmers University of Technology, Göteborg, Sweden

M. Johansson
Department of Technology, Management and Economics, Chalmers University of Technology, Göteborg, Sweden

© The Author(s) 2022
Z. S. Klos et al. (eds.), *Towards a Sustainable Future - Life Cycle Management*,
https://doi.org/10.1007/978-3-030-77127-0_18

the original material value via reduction in industrial waste volumes and decrease of the total virgin raw material production per one unit of output, in addition to increasing the homogeneity of wasted material with better waste segregation [5]. The latter enables moving from landfill and waste incineration towards recycling, remanufacturing, reuse and repair (reverse material flow).

The importance of the production phase in the value chain is essential in sustainable development and circular economy as it currently accounts for 33% of total global energy consumption and 38% of direct and indirect carbon dioxide emission [6]. In addition, the production phase contributes to different environmental effects including increased (virgin) raw material and energy consumption, great industrial waste volumes and airborne emissions.

The automotive industry is of particular interest to study, due to the fact that it negatively contributes to the majority of environmental effects. According to the European Automobile Manufacturers Association [7], the production phase in automotive industry in 2017 contributed to 38.8 million MWh energy consumption, 9.47 million-ton CO_2 emission, 56.89 million cubic metre water consumption, 1.4 million-ton waste generation and 38.6 thousand-tons of volatile organic compounds emission. Considering material flows, automotive industry is of interest since metal is used as the primary product material, while several other material fractions such as plastics, chemicals, cardboard, wood and combustible are consumed as auxiliary materials. Furthermore, the generated waste from automotive industry are common residuals mainly including scraped aluminium and steel, chemicals and hazardous waste and packaging materials such as plastics, cardboard, wood and combustible waste. Figure 1 shows the common material flows in automotive industry using a framework presented by [8].

This paper presents and compares the current flows of metals and plastics in the automotive value chain by two criteria, namely, information flow and standards and regulations. An underlying reason is to learn from the relatively better working metal recycling when improving plastic recycling and highlight common needs in both loops. This contributes to the material circular flow knowledge by pinpointing

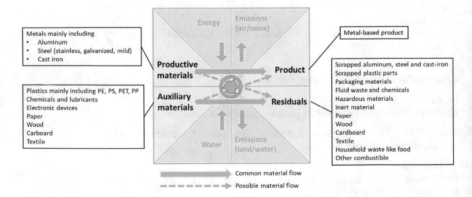

Fig. 1 Common material flows in automotive industry

the gaps, similarities and differences of two material flows as well as extending the collaboration in recycling loops. It is also a help for improving the overall material efficiency and industrial waste management practice.

2 Research Methodology

Research presented in this paper was carried out as a part of an ongoing Swedish research project called "Circular Models for Mixed and multi Material Recycling in manufacturing extended Loops" (CiMMRcc), and with an extension pre-study on plastic loops in a research called "Sustainable plastic use by managing uncertainties for the market actors". The project aims to explore opportunities for extended collaboration in recycling loops, especially studying knowledge transfer, information flows, incentives, standards and regulations and business models for improved material recycling, and contributes to the area of circular economy [9] and sustainable supply chains [10]. With limited understanding and lack of empirical studies on characteristics of metal and plastic flows in an automotive value chain, a case study methodology was adopted to fulfil the research objective, consisting of real-time empirical data from different companies within the automotive value chain and a limited literature review. The studied companies are all value chain actors within the automotive industry but in the two separated metal and plastic loops. Studied companies range from primary production of raw materials, product manufacturers, foundry and waste management entrepreneurs to recycling companies.

Although the metal and plastic flows are generally different, the information flows and communication, incentives, business models and standards and regulations for these flows should not differ to a very large extent in order to have a successful recycling flow. Lack of recycling initiatives in any of these flows causes losing material values captured during the linear production processes of materials and products (linear production process as opposed to reverse processes of reusing, repairing, remanufacturing and recycling). As a result, multiple case design with embedded unit of analysis [11] was used, where one case represents metal value chain and the other represents plastic value chain (see Fig. 2). The product manufacturers in both cases are multinational manufacturing companies with global footprints in the automotive industry that use metals as primary production material (productive material) and plastics as auxiliary materials (see [5] for definitions). The selection of companies was mainly based on their close collaboration and project connections, which in turn was primarily based on their enthusiasm in improving their current systems for achieving sustainability and circularity in their materials flows. This close co-research connection facilitated accessing and data collection, arranging semi-structured interviews [12], direct observation by visiting operation sites [11], reviewing relevant documents and monitoring material and waste flows. In the first set of interviews, a total of eight people was interviewed, although some (waste management entrepreneurs) answered two sets of questions related to both metal and plastics. Each semi-structured interview lasted between 30 and 90

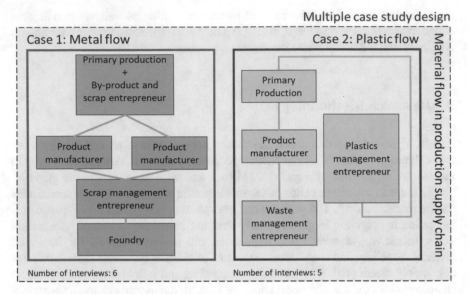

Fig. 2 Case study design

minutes and incorporated predefined questions regarding metal and plastic flows in value chain with several criteria such as information flow, regulation and business models. A second set of interviews included four interviews with six people from the same plastic flows as the first set of interviews. Considering these ongoing market changes, the supplier - user requirements were further elaborated. Data analysis and interpretation was performed within a very short time interval after data collection, as suggested by [11]. Consistency between interviews and for both material flows was maintained throughout the data collection and analysis, by continuously reviewing, comparing and discussing the results with project members including practitioners from the studied companies.

3   Empirical Findings and Discussions

The empirical findings and following discussions presented in this section are based on performed interviews of actors in the value chain shown in Fig. 2, reviewed documents and also direct observation in operation sites (where possible). This section is divided into the main material flows in automotive industry, i.e. metal and plastic. For each material flow, the two main criteria, i.e. information flow and regulations and standards, are discussed.

3.1 Metal Flow

Several different types of information and data are communicated between different actors within the value chain. However, our focus was on information that helps circulating the metal flow (mainly metal scrap in order to close the loop) for recycling and reuse. That being said, the main information flow within this value chain includes material type and fraction, sorting degree, physical shape and dimension, amount in terms of weight in kg, chemical composition and price. There has been a general consensus among the actors (interviewees) that currently sufficient amount and type of information is available (e.g. exact chemical composition of the waste), and there is no need to dig deeper to find the information. However, the problem is mainly information sharing, communication and transparency. It is also the matter of actors' ambitions to ask for more information and to put more effort and time in obtaining necessary information and analyse them for improvement. For instance, the communication between the scrap management entrepreneur and product manufacturers (and also right department, in particular purchasing who buys materials) could be improved; in a specific example, changing the material and/or supplier of components was not clearly communicated with scrap management entrepreneur. The main reason for this was that the product manufacturers were not aware that changing alloy or chemical content of materials and components would have serious consequential effects in end-of-life management and recycling. This issue does not require any regulation or legal intervention, but better information sharing and communication between the actors. Another issue related to information is variability. The majority of metal scraps and waste are generated due to deviations, errors and mistakes in production (see also [8, 13]); therefore, types, physical shapes and weights differ significantly from one to another. This variation negatively affects the number of transportations where sometimes half-full trucks are transporting the waste. There have been some unsuccessful attempts to solve this issue such as using sensors in the metal bin, but it did not work as good as for fluids. In another example, a camera was placed to monitor the content of the metal bin, but sharing this type of data between companies was problematic due to IT regulations. Nevertheless, it could be concluded that improvement actions should start from the product manufacturer, for instance, with better sorting or better communication of information with other actors.

Taking regulation and standards into consideration, there was an agreement among the actors that quality standards for secondary material (metal) would not only ease pricing based on value but also help improve waste segregation and recycling. However, there was also consensus that forced additional standards may disturb the market and distort the competition. The metal primary production actors believed that having more standardized fractions would lead to more complexity and therefore more cost would relate to type of scrap, handling systems and storage. According to metal primary production actors, European standards bring difficulties due to import and export regulations between different countries which take a lot of time and knowledge to fulfil those requirements. The interviewee from a foundry company also asserts "I don't see any need for additional standards on iron and steel, but how well one manages to follow the standards is important … we don't need any further pressure or temptation". In Sweden, companies also follow the national iron standard (Järnbok), which does not always align with standards from other countries, e.g. when buying iron from Germany. Hence, in the long term, an international iron standard is needed to facilitate recycling. There was also difference of opinions on whether regulations and standards should be material or industry specific.

To summarize our empirical results on metal, information flow, actors' role, technology development, market, regulation and standards, product design and behaviours work quite fine with the current infrastructure of metal flow, although several minor improvements (such as given in the examples above) can be made.

3.2   Plastic Flow

The main information flow within the reverse plastic value chain (mainly recycling and reusing) includes plastic type, fraction and prime material, sorting degree and cleanness, shape and dimension, volume in terms of weight in kg, chemical composition and price. Unlike the metal flow, the general consensus among the actors was that more and better information and communication are needed, particularly on exact sorting degree and exact type of plastic and fraction, including details on risk of contamination with unwanted substances. The information flow from the plastic supplier to product manufacturer seems to be working better than the information flow to the waste management and also further back to the plastic management entrepreneur (see Fig. 2). In spite of this, also the information required and given from the supplier has gaps. For instance, it is now the product manufacturer that almost solely decides on the selection of supplier and also type and material of the plastic packaging of purchased components. This decision is mainly based on requirements on the products' protection during transport, due to legal issues (the one who determines the packaging is responsible for parts broken during transport), and until just recently, the footprint of the packaging material has not been in requirements. However, such decisions could involve waste management entrepreneur to explore and discuss opportunities to exclude plastic packaging to a certain possible level and use less additive to ease recycling.

According to the interviews with actors in the plastic value chain, there are several issues with the plastic recycling, including the following:

(1) Recycled plastic does not always have the exact same quality/properties as specified in current parts.
(2) Price of recycled plastic has often been more expensive compared with the relative low prices of plastics made of virgin material, although recently virgin prices have been perceived as more volatile according to the second sets of interviews.

(3) The reverse value chain is not as smooth and steady as the forward value chain and has lots of interruptions, delays and bottlenecks due to unevenness of availability of recycled plastics and variable lead time in collection of plastic waste and recycling. Within the automotive industry, manufacturing companies have the obligation to produce the exact same product for several years, e.g. 10 years, and hence, they need a guarantee that the recycled plastic with the same properties and quality is available for the next 10 years and can be delivered steadily in order to be able to produce the same product with the same properties and quality.

(4) There has not been a customer requirement on the share of recycled plastic in the products. Increasing the share of recycled plastic without the customers' requirement and with current higher prices of recycled plastic compared to virgin plastic would make the product more expensive and hence less competitive.

(5) The interviewees also highlighted issues with the plastic recycling process itself, including lack of plastic sorting. Increase in the number of bins to better segregate plastics into more fractions is a great challenge because usually there is not enough space inside and outside the factories. In addition, managing five to eight different plastic fractions would be time-consuming and expensive for the product manufacturer considering the relatively low market prices. There are also more combustible bins on the shop floor with less walking distance than a specific plastics bin. Consequently, with intrinsic indolence of human being and weariness and exhaustion from work, plastics are usually discarded in combustible bins. One potential solution would be to somehow achieve higher market price for the sorted recycled plastics.

(6) Unlike the household plastic waste that is separated after collection by the waste management entrepreneur in exchange of a small fee, in the industrial system, the product manufacturer is not willing to pay the waste management entrepreneur for segregation, which substantially limits the segregation. At the same time, factory workers do not understand the need for sorting plastics in multiple fractions as just one bin for plastics is used for households. Therefore, a behavioural change or education/training in industry is needed for further waste segregation of plastics.

(7) Low volume fractions are not economically viable for separation and recycling. According to the interviewees and our previously published study [14], polyethylene (PE) account for 40–74% of total plastic waste from automotive manufacturing, which can and must be separately segregated for recycling. However, the remaining fractions (such as polypropylene – PP) have relatively low volumes, and hence, efforts for separation are perceived not to be economically viable.

(8) There is a transportation efficiency issue with correlated high costs that trucks need to be full for economic and environmental reasons. A sufficient volume for each transport can be 3–4 tons for PA (polyamide) and 5 tons for PP, a relatively high amount compared to the general low volumes of sorted plastic waste in many automotive plants.

(9) Separation should be based on polymer which is difficult for operators to distinguish the type of plastic; hence, environmental education as well as plastic labelling is important as unmarked plastics cannot be segregated.

(10) Segregated plastics should not be contaminated with dirt, sand, metal chips, etc.

(11) There is a lack of information, e.g. precise volume, sorting degree and type of material for transportation. Not all companies provide the necessary information to the waste management or plastic management entrepreneur. Sometimes, the information provided is also wrong. Therefore, extra time and cost have to be put in testing the fractions randomly by the waste management or plastic management entrepreneur.

(12) Current technologies for plastic segregation and recycling (e.g. segregation machine based on plastics colour shade) are inefficient and expensive, and also the process is time-consuming, which neither the customer nor the product manufacturer willing to pay for that.

(13) Demand for recycled plastics has been low and separation is being done manually; hence, there is a high associated cost.

(14) It is simply too expensive to recycle plastics, compared to incinerating it. However, this issue is related to Sweden where it is relatively cheap to incinerate to produce household heat; hence, little incentive exists for industry to recycle more. Government intervention or tax is needed to solve this problem and gives motivation to make changes, for example, by looking into other countries such as France where it is rather expensive to incinerate or the Netherlands where it is forbidden to incinerate certain materials.

Taking regulation and standards into consideration, in general it was believed that more regulation would be helpful to close the plastic loops; however, the so-called carrot approach was more favourable than the stick approach. During the interviews, several regulation suggestions were proposed including the following:

• Better suited industrial waste fractions standards (not necessarily regulated), adapted for how to sort to reach marketable fractions and material properties.

• Regulations and standards that take away tax on recycled material to lower costs for using recycled plastics. Maybe also subsidies to start demand for recycled plastics will help. Likewise, shifting tax from labour to tax on virgin materials might help sort and recycle plastics better.

• Regulations and standards on having the same type of plastic for all packaging to reduce diversity and ease sorting. Purchasers can make demands on suppliers to use only a certain type of plastic.

• Regulations and standards on number of polymers allowed in a single product. Many products include several types of plastics which are difficult to separate. Shredding or incinerating those products is the only current possibility. Perhaps some legislation on not mixing several types of plastics might be helpful.

• Regulations and standards on labelling the plastics. Unmarked plastics cannot be segregated into plastic fraction and hence are thrown in combustible bins without any recycling. Companies could demand suppliers to mark their plastics.

Although label is mainly for end-customers, it might lead to OEM wanting a higher share of recycled materials in their parts.

- Regulations and standards to force product manufacturing companies to take responsibilities for their plastic waste and segregate it (e.g. PE as mentioned earlier).
- Tax on waste incineration; alternatively, prohibiting incineration of recyclable materials.
- Regulations and standards to put requirements on sorting and recycling waste; alternatively, tax on unsorted waste.
- Regulations and standards to put requirements for manufacturers to use a certain level of recycled material.

Nevertheless, some concerns regarding regulations were also expressed including limiting regulation from European Union that hinder the plastic recycler and recycled plastic seller to purchase and import from non-EU countries, which exacerbate the abovementioned issue of insufficient volume. It was of concern that having strict legal requirements only in Sweden might lead to a shift to other countries outside Sweden to stay competitive in the market; therefore, regulations and standards must aim at EU and/or global level. Furthermore, waste management entrepreneurs were concerned about standardization that would also mean increased logistics and increased requirements of more bins and space. Plastics have a large volume compared to weight. Therefore, for efficiency transportation, a shredder is needed to make plastic more compact to increase the volume for each transportation.

There was difference of opinions on whether regulations and standards should be material or industry specific. One example of industry-specific regulations and standards was to have a simple guideline for automotive industry to pinpoint few possible improvement steps for better plastic segregation and recycling. An example of material-specific regulations and standards was to put tax on certain virgin materials. However, this proposition was argued to be counterproductive in a way that it might decrease the use of virgin plastic but not necessarily increase the recycled plastics. Tax cut could improve the situation, but the price of recycled plastic is much higher than the tax on it and therefore would only have a very limited effect.

There is some sort of circular business model in the studied product manufacturing company to reuse some plastic components where slightly lower properties are required and also some variations are possible. Nevertheless, proper reuse and remanufacturing of plastic parts is not possible. There is not much commodity between parts and it is much easier to melt down plastic and recycle it. However, it would be still very costly to have an additional flow of used plastic parts in production. This requires a big design change in the automotive industry, e.g. less durability requirement in vehicles for carpooling.

4   Conclusion

There has been a consensus among interviewees that competition for recycled material will increase and more manufacturing companies will ask for recycled material. Hence, waste management need to be integrated in daily operations, to effectively meet the increased demand. According to our empirical study and performed interviews, metal waste is segregated to a high degree and with low level of errors, while mostly the exact chemical composition of the metal scrap is known. For instance, to get the best recycling option, steel is not mixed with non-ferrous metals like aluminium or copper. The demand for recycled metals is also relatively good and current standards are fine. However, there are still some improvement potentials in metal flow management such as better communication and information sharing among actors which could positively affect the number of transportations and incoming material selection for better recycling options at the end-of-life. These issues are apparent also in the small plastic recycling flows. On the other hand, the major problem for plastic recycling is that plastic waste has low level of segregation with high level of errors in the segregation process. The full chemical composition is usually not known either. As a result, the plastic waste needs to be regularly checked, which implies additional waste handling and administration. With such low level of separation (due to several reasons discussed earlier) and correlated low volumes, inefficient transportation, quality errors and contaminations, technological issues and top of all insufficient demand for recycled plastics and low price of virgin plastics, recycling were commonly not regarded as economically interesting for companies in the value chain. There is a rather great requirement for more standardized fractions, and legal requirement as well as an economic or regulatory motivation.

As it can be perceived from literature and our empirical study among actors in the value chain, the metal flow is more matured than the plastic flow. This can be argued with the long history of metal industry development since the 1850s, and even far back earlier in the prehistory where human used metal to build tools and weapons. On the other hand, plastic industry development is relatively new, started in almost the 1950s. While the plastic manufacturing and use in a variety of applications expanded exponentially, little thought and research has been given to the impact of such quick growth and to develop proper waste management system for plastics. In addition, this can be reasoned with the fact that the metallurgical properties of metals allow them to be recycled repeatedly with no or neglectable degradation in performance and quality, and from one product to another. Deteriorating, plastic recycling is challenging, thanks to the variety of additives and blends used in manufacturing, low demand of recycled plastics and cheap price of virgin plastic.

With such underdeveloped plastic waste management and the sudden decision of China in 2016 to terminate importing plastic waste for recycling, we need to create the motivation in developed countries to develop an effective domestic recycling infrastructure, expand domestic market for recycled plastics, change the product design for better recycling and reuse and make the business model economically more interesting for actors in the value chain. A developed market and competition can be enablers for self-imposing regulation in increasing the share of recycled material in the products, increasing tax on virgin materials and reducing tax on recycled materials, subsidies, etc., which will happen gradually and naturally over time.

Our studies were carried out in automotive industry where metal is the dominant material, and circulation (recycling in this case) of the dominant materials is of most importance due to volume and value. However, this should not justify the low circulation/recycling rate of other materials, particularly plastics.

References

1. Litos, L., & Evans, S. (2015). Maturity grid development for energy and resource efficiency management in factories and early findings from its application. *Journal of Industrial and Production Engineering, 32*(1), 37–54.
2. Mistra Closing the Loop, Closing the Loop for industrial by-products, residuals and waste: From waste to resource. 2015, The Swedish Foundation For Strategic Environmental Research, Sweden.
3. Worrell, E., Allwood, J., & Gutowski, T. (2016). The role of material efficiency in environmental stewardship. *Annual Review of Environment and Resources, 41*(1), 575–598.
4. Ellen MacArthur Foundation, Circularity indicators: an approach to measuring circularity-Methodology. 2015, Ellen MacArthur Foundation, UK.
5. Shahbazi, S. (2018). Sustainable manufacturing through material efficiency management. In *Innovation, design and engineering*. Mälardalen University: Sweden.
6. Garetti, M., & Taisch, M. (2011). Sustainable manufacturing: Trends and research challenges. *Production Planning & Control, 23*(2–3), 83–104.
7. The European Automobile Manufacturers' Association, A., The automobile industry pocket guide 2018/2019. 2018.
8. Shahbazi, S., et al. (2018). Material efficiency measurements in manufacturing: Swedish case studies. *Journal of Cleaner Production, 181*, 17–32.
9. Ellen MacArthur Foundation, Towards the circular economy. Economic and business rationale for an accelerated transition – Executive Summary, 2012. Ellen MacArthur Foundation.
10. Brandenburg, M., et al. (2014). Quantitative models for sustainable supply chain management: Developments and directions. *European Journal of Operational Research, 233*(2), 299–312.
11. Yin, R. K. (2014). *Case study research: Design and methods* (5th ed.). Sage.
12. Kvale, S., & Brinkmann, S. (2009). *InterViews: Learning the craft of qualitative research interviewing*. Sage.
13. Kurdve, M., van Loon, P., & Johansson, M.. (2018). *Cost and value drivers in circular material flow logistics*. In 5th International EurOMA sustainable operations and supply chains forum.
14. Shahbazi, S., et al. (2016). Material efficiency in manufacturing: Swedish evidence on potential, barriers and strategies. *Journal of Cleaner Production, 127*, 438–450.

Open Access   This chapter is licensed under the terms of the Creative Commons Attribution 4.0 International License (http://creativecommons.org/licenses/by/4.0/), which permits use, sharing, adaptation, distribution and reproduction in any medium or format, as long as you give appropriate credit to the original author(s) and the source, provide a link to the Creative Commons license and indicate if changes were made.

   The images or other third party material in this chapter are included in the chapter's Creative Commons license, unless indicated otherwise in a credit line to the material. If material is not included in the chapter's Creative Commons license and your intended use is not permitted by statutory regulation or exceeds the permitted use, you will need to obtain permission directly from the copyright holder.

Social Life Cycle Indicators Towards a Sustainability Label of a Natural Stone for Coverings

Elisabetta Palumbo and Marzia Traverso

Abstract The stone industry plays an important economic role in Italy as well as worldwide, and its products are part of the construction sector for hard coverings. The relevance of these products led the European Commission to develop specific criteria for natural stone within the Ecolabel scheme for hard coverings. In order to provide environmental information and to establish and maintain their comparability, the eco-labelling schemes recognized the life cycle assessment (LCA) as a scientific method to be employed when describing the environmental performance of the products. In its current form, the European Ecolabel scheme only considers environmental impacts and overlooks significant social impacts, especially for the category of stakeholders most affected during the extraction and manufacturing phases: workers. The main purpose of this study is to define a set of social criteria to be added to the revised version of the European Ecolabel with reference to issues concerning natural stone covering products. In particular, according to the updated guidelines for the social life cycle assessment by UNEP/SETAC Life Cycle Initiative (2019), we have identified that the "health and safety" impact category as it relates to workers during the extraction and manufacturing phases of the products must be considered a priority. The results provide a set of criteria for the S-LCA inventory which should be added to the Ecolabel guidelines when assessing the natural stone covering sector. Integration of the social sphere with the results obtained from the LCA study would provide reliable and more complete information on the sustainability of the natural stone product.

This represents a first step towards the inclusion of similar criteria for other covering products.

E. Palumbo (✉) · M. Traverso
RWTH Aachen University, Institute of Sustainability in Civil Engineering (INaB),
Aachen, Germany
e-mail: elisabetta.palumbo@inab.rwth-aachen.de

© The Author(s) 2022
Z. S. Klos et al. (eds.), *Towards a Sustainable Future - Life Cycle Management*,
https://doi.org/10.1007/978-3-030-77127-0_19

1   Introduction

The stone industry plays an important economic role in Italy and worldwide. In fact, the stone and marble industry is a sector that in certain geographical areas contributes to the local production and employment capacity.

In the global trade of natural stone (marble, granite, stone, travertine) in 2015, Italy ranked second worldwide (13.5%) after China, which holds the largest market share with 35.8% (Japan and other countries in the region are among its most important partners) (Table 1) [1, 2]. Italy, with its production areas covering highly specialized activities and extracted rock types, still plays a strategic role in the production and exportation of stone materials. In 2018, marble, travertine and alabaster products achieved high exports of around 402,685 tonnes [3, 4].

Natural stone is widely employed in the building and construction sector, in particular as a wall cladding material due to its attractiveness, durability and versatility [5].

Nevertheless, this sector has a negative impact on the environment and society as a result of the large amount of waste generated by extraction and processing (30–50% of the extracted gross quantity) [6], dust pollution linked to the extraction process and water pollution caused by cutting processes [7].

By the twentieth century, the location of mining sites had shifted from developed to developing countries, with two important consequences: firstly, the provision of less expensive raw materials from non-European Union countries led Europe to rely more on imports; secondly, the environmental and social impacts shifted to countries that are major producers where attention to sustainability issues is lacking, making sustainability assessment necessary.

The interest in social and ethical issues raised by a product along its life cycle is increasing, particularly in sectors such as raw material extraction and mining where there are potentially high health and safety risks for workers.

As far as natural stone is concerned, the Italian ornamental stone industry is one of the main producers worldwide.

In Italy, in 2015 alone, approximately 5.3 million tonnes of ornamental stone were produced; the regions with the highest number of quarries (20 or more) are Tuscany, Lombardy, Apulia and Veneto [8]. The quarries of Carrara in Tuscany, for

Table 1 Quarry productions and processing wastes in the world (readapted by [3])

Leading stone countries	Quarry production	Processing waste	
	(kt)	(kt)	(%)
China	45,000	22,768	50.6
India	21,000	6285	29.9
Turkey	10,500	2493	23.7
Brazil	8200	2990	36.5
Italy	6500	2485	38.2
Spain	4750	1641	34.5
Portugal	2700	812	30.1

example, provide most of the marble used in Italy and Europe for sculpture and other ornamental work, along with a large number of blocks, which are sent in raw or finished form to all parts of the world [9].

Given the importance of this sector, the social impact issue cannot be ignored. Data from the Italian National Institute for the Prevention of Accidents at Work (INAIL) [10] shows that the number of accidents and occupational diseases in the "Quarrying of ornamental and building stone, limestone, gypsum, chalk and slate (NACE 08.11) sector" is not insignificant.

Starting with statistical data collected on accidents at work in this sector from the INAIL database, this study aims to highlight the integration of social aspects of sustainability regarding natural stone within the Ecolabel scheme (ISO 14024:2018) into the current revision of the criteria for the Ecolabels of hard floor coverings (Commission Decision 2009/607/EC).

The main goal of this study is to identify the social hotspots and social impacts that should be added as assessment criteria in the revised Ecolabel scheme for natural stone coverings.

In order to achieve the above-mentioned aim, this study is divided into three parts:

(1) Background of the social criteria considered in official documents, in literature and in the existing Ecolabel schemes (e.g. European Ecoflower) with particular attention to stone and hard surfacing and the field limitations of this study
(2) Identification of the weaknesses of the natural stone sector as regards health risks and injuries to workers during quarrying and manufacturing processes, based on a review of the literature on work medicine and a survey of the statistical data relating to workers' health – taking the database developed by the Italian National Institute for the Prevention of Accidents at Work (INAIL) as a reference and based on an investigation of the Social Hotspots Database (SHDB), which provides social risk data at sector and country level, focusing on the global risk to health and safety in both stone quarrying and manufacturing processes
(3) Proposal of a set of criteria for S-LCA inventory for natural stone coverings

The social indicator set developed can serve both as a proposal for the Ecolabel criteria revision with a view to social considerations and as a guide on how to determine the sustainability performance of the hard coverings. Furthermore, a list of challenges and benefits for social life cycle assessment (S-LCA) implementation can be identified and presented to support the current revision of the Guidelines on Social Life Cycle Assessment [11].

2 Aims of the Study and Assumptions

The main reference in the social life cycle assessment (S-LCA) is represented by two important guidelines: those developed by UNEP [11] which define the S-LCA as a complementary method of life cycle assessment (ISO 14040, 2006) and the

Handbook for Product Social Impact Assessment [12], which was developed over 3 years of work by the Roundtable for Product Social Metrics. Both methodologies – the second derived from the first – identify the main stakeholder groups: workers, users/consumers and local communities. For each of them, a set of impact categories and its relative indictors was proposed. According to the UNEP guidelines, few case studies can be identified, and one of the first concerns natural stone products [13, 14].

The literature review conducted some years ago by Hosseinijou et al. [15] on the integration of social aspects into a life cycle format for building materials counts nine papers as the most remarkable: O'Brien et al. (1996), Schmidt et al. (2004), Dreyer et al. (2006), Hunkeler (2006), Norris (2006), Weidema (2006), Reitinger et al. (2011), and Lagarde and Macombe (2013) and Jørgensen et al. (2008), which reviewed most of the current S-LCA literature.

Based on an overview of the social aspects identified in 12 major S-LCA sources of the literature, and in accordance with the social impact categories proposed in the UNEP guidelines, Siebert A. et al. [16] in a recent study applicable to wood-based production systems in Germany identified a set of 15 social aspects. Of these 15 aspects identified, it was estimated that the most used indicators in the 12 case studies are discrimination/equal opportunities, fair salary/wages, health conditions/health and safety, freedom of association and collective bargaining (Fig. 1).

Social impacts in the mining sector appear to have been discussed for over 10 years. Mancini et al. [17, 18] deal with this type of problem by combining the Social Hotspots Database (SHDB), a global database that eases the data collection burden in S-LCA studies [19], with the social impacts in the mining sector documented in 12 references (9 scientific papers and 4 reports from international organizations). The SHDB, following the UNEP S-LCA guidelines, indicates the social risk of the main countries and sectors in the world. Not all the data from impact subcategories is contained in the SHDB, but there is enough to provide a good overview. The study divides the social impacts into positive and negative and checks which impacts are included in the Social Hotspots Database. Therefore, as a first step of the research, taking into account all the impacts considered, we selected only the negative ones dealt with by both multiple sources of literature and the SHDB, and specifically "negative health and safety impacts on workers" and "environmental impacts affecting social conditions and health".

The Global Ecolabelling Network (GEN), established in 1995, is a non-profit association of Type I Ecolabel organizations and has members in several countries. To improve, promote and develop the Ecolabels of products and services globally and to enhance mutual trust and recognition among various reputable Type I Ecolabelling programmes in accordance with ISO 14024, the GENIUS framework was developed which, in addition to verifying that each programme "abides by ISO 14024 principles and is robust and trustworthy, the process can inspire your employees around a shared societal goal" (The Global Ecolabelling Network, 2017).

An analysis focused on the ecolabelling programme for hard coverings within GEN showed that a very small percentage of schemes evaluate the social issues and adopt social indicators related to the health and safety of workers.

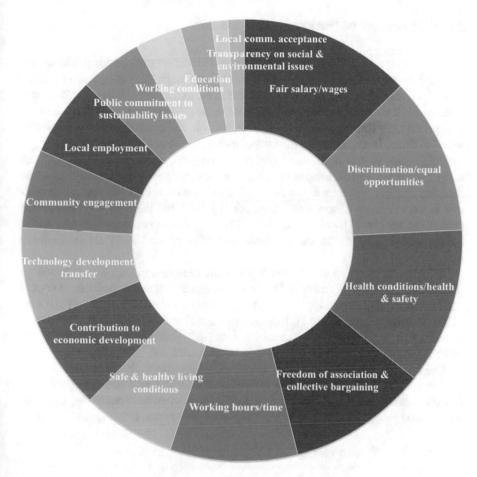

Fig. 1 Set of social aspects applied in S-LCA case studies identified by [12]

One of the most pertinent Ecolabel schemes in this sense is Australia's Good Environmental Choice Australia (GECA) for hard surfacing, which with the introduction of Section 10 on "social and legal requirements" includes criteria linked to aspects such as equal opportunities and the safety and protection of workers.

In light of the above and in accordance with the stakeholder categories and subcategories suggested by UNEP "Guidelines for Social Life Cycle Assessment of Products", this study focuses on workers' health and safety: "negative health and safety impacts on workers".

3   Weaknesses of Natural Stone Sector

The natural stone extraction, transportation and manufacturing sector produces relevant environmental, social and economic impacts internally, locally and globally.

Guidelines for the safety of human health in the extraction industries were developed by the European Commission in Directive 2006/21/EC together with measures and procedures to reduce any adverse effects on the environment (in particular water, air, soil, fauna, flora and the landscape) within waste management.

References in literature show that non-European stone quarrying processes release elements into the environment such as dust, sludge or other industrial waste that may be toxic and constitute a health risk to humans: substances that are hazardous to the cardiorespiratory system, physical fitness and the body as measured at stone quarries [20, 21], pulmonary problems [22], skin dermatoses [23] and ocular health hazards [24], and in general the health of employees and their productivity and efficiency [25].

An analysis of occupational accidents in the mining sector in Spain, based on data from the Spanish Ministry of Employment and Social Safety between 2005 and 2015, shows that the most typical accidents are body movement involving physical effort or overexertion and, in underground mines, fractures, slips, falls or collapse. Moreover, it highlights that the lack of safety education and training is one of the most influential factors leading to mining injuries [26].

The INAIL database on reported work-related injuries in the quarrying of ornamental and building stone sector in Italy shows a fairly stable trend. In particular, this data shows that in the last 4 years, accidents at work have decreased by about 10%, while professional illnesses have increased by approx. 6% (Fig. 2).

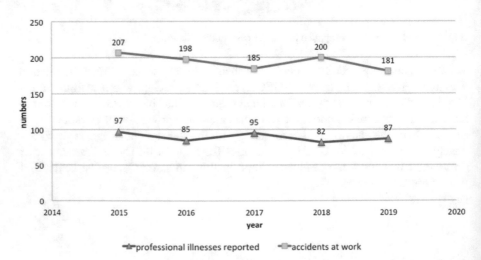

Fig. 2 Numbers of professional illnesses reported and accidents at work in the extraction of ornamental stone sector from 2015 to 2019 in Italy (elaborated by the authors from [10])

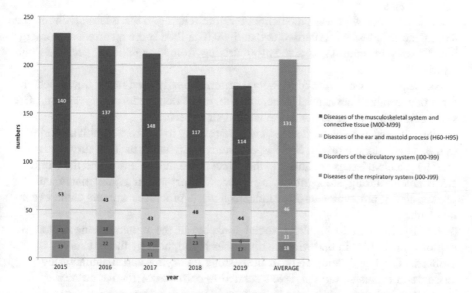

Fig. 3 Number of workers with the diseases classified according to ICD-10 from 2015 to 2019 in Italy (elaborated by the authors from [10])

A more detailed analysis of the professional diseases classified according to the International Statistical Classification of Diseases and Related Health Problems, version 2010 (ICD-10), indicates that the four main burdens of disease are respectively diseases of the musculoskeletal system and connective tissue (M00-M99), diseases of the ear and mastoid process (H60-H95), disorders of the circulatory system (I00-I99) and diseases of the respiratory system (J00-J99). Specifically, Fig. 3 shows the number of workers with the diseases recorded from 2015 to 2019 and the average for each of the four main illness types.

4 Outcomes

This study, which aimed to identify a set of social criteria to be added to the revised version of the European Ecolabel for natural stone covering products, has identified critical issues related to the social dimension through the following steps.

Starting with the screening of the five main stakeholder category groups (workers/employees, local community, society, consumers and value chain actors) to be considered in the social impact assessment in accordance with the UNEP guidelines (2009) and the revised version (2020) [11], we identified the priority of taking into account the health and safety aspects of workers who seem to be the most affected due to the intrinsic risks of the activities they perform during the extraction and manufacturing phases and their exposure to dust.

An initial review of work medicine literature relating to the issues arising in the natural stone industry was carried out, and we identified some recurrent and emerging diseases in addition to discomfort arising from occupational accidents and injuries.

Subsequently, we reviewed the social criteria already used in the flower scheme for products other than natural stone, and the results obtained from surveys on LCA studies filled in by companies in the natural stone sector.

We collected and analysed statistical data relating to workers' health and injuries in the natural stone industry, limiting our study to Italian data. This survey showed that the principal issues are linked to the effect of the dust released into the workers' environment during stone quarrying processes or within stone manufacturing phases, sludge production or other industrial waste processes workers come in contact with.

Finally, in order to highlight social hotspots in the mineral stone sector, we explored the SHDB in line with the outcomes highlighted by the last survey on the global risk to health and safety in both stone quarrying and manufacturing processes, and evaluated the risk levels related to chronic obstructive pulmonary disease due to airborne particulates in the workplace.

In conclusion, considering the results produced by this investigation from both work medicine literature and a survey of statistical data from the National Institute for the Prevention of Accidents at Work (INAIL), the main impacts are due to:

- Dust emission with consequences for pulmonary and cardiorespiratory functions, as well as dermatologic and ocular diseases
- The risk of accidents at work
- The risk of accidents at work caused by contact with water and sludge which may be harmful to human health

In addition to these aspects, the outcomes from the INAIL statistics database show that the major cause of accidents is movements in the workplace that can result in muscular problems (Fig. 3).

On the basis of these observations, it is important to define and integrate social criteria related to workers' health and safety in the natural stone coverings industry, to be added to the Ecolabel of these products. This would provide reliable and more complete information on their sustainable performance, as a first step towards the inclusion of similar criteria for other covering products.

5   Conclusion and Recommendations

These studies reveal the strong association between the environmental and social dimensions of the manufacturing processes. While the environmental dimension has been broached by voluntary methods to certify and label environmental performances, such as the Type 1 label (Ecolabel), social aspects were left out. Furthermore, no consideration was given to the fact that data and indicators to estimate local environmental impacts can also support the assessment of the social impacts related to health and safety. It should be noted that national regulations on the health and safety of workers are in place, but they are not included in product labelling.

The study we conducted shows that Type I Ecolabel statements should contain a more complete assessment and documentation of product sustainability. Our suggestion is that the inclusion of social criteria in the Ecolabel scheme is clearly necessary to avoid an incomplete assessment of the impact of the natural stone manufacturing process.

This work can be considered a first step in the process of identifying a set of social criteria related to the workers' stakeholder category. The limitations of the study lie in the fact that we only analysed one of the important stakeholders closely involved in the social issue.

Therefore, future work should broaden the field of analysis for this proposal and investigation, first and foremost to the impacts of subcategories on "local communities".

References

1. Montani, C. (2016). *XXVII World marble and stones report 2016*. Aldus Casa di Edizioni in Carrara.
2. U.S. Geological Survey, Mineral Commodity Summaries 2019. (2019). U.S. Geological Survey, 200 p. https://doi.org/10.3133/70202434.
3. https://stonenews.eu/italys-natural-stone-products-exports-2018/
4. https://www.intracen.org/
5. Ferreira, C., Silva, A., de Brito, J., Dias, I. S., & Flores Colen, I. (2021). Definition of a condition-based model for natural stone claddings. *Journal of Building Engineering, 33*, 101643. https://doi.org/10.1016/j.jobe.2020.101643
6. Marras, G., Bortolussi, A., Peretti, R., & Careddu, N. (2017). Characterization methodology for re-using marble slurry in industrial applications. *Energy Procedia, 125*, 656–665.
7. Abu Hanieh, A., AbdElall, S., & Hasan, A. (2014). Sustainable development of stone and marble sector in Palestine. *Journal of Cleaner Production, 84*, 581–588. https://doi.org/10.1016/j.jclepro.2013.10.045
8. Zanchini E., e Nanni G. (2017). *Legambiente – Rapporto cave*, GF pubblicità – Grafiche Faioli.
9. Primavori, P. (2015). Carrara marble: A nomination for 'Global Heritage Stone Resource' from Italy. *Geological Society London Special Publications, 407*(1), 137–154.
10. https://bancadaticsa.inail.it/bancadaticsa/login.asp (Accessed 01.08.2019).
11. UNEP, Guidelines for Social Life Cycle Assessment of Products and Organizations 2020. Benoît Norris, C., Traverso, M., Neugebauer, S., Ekener, E., Schaubroeck, T., Russo Garrido, S., Berger, M., Valdivia, S., Lehmann, A., Finkbeiner, M., Arcese, G. (eds.). United Nations Environment Programme (UNEP), 2020
12. Fontes, J. et al. *Handbook of product social impact assessment version 3.0*, 2016. https://product-social-imocat-assessment.com (Accessed 02.08.2019).
13. UNEP/SETAC Life Cycle Initiative. (2011). *Towards a life cycle sustainability assessment-making informed choices on products*. Paris: United Nations Environment Programme.
14. Finkbeiner, M., Schau, E. M., Lehmann, A., & Traverso, M. (2010). Towards life cycle sustainability assessment. *Sustainability, 2*, 3309–3322. https://www.mdpi.com/2071-1050/2/10/3309/pdf

15. Hosseinijou, S., Mansour, S., & Shirazi, M. (2014). Social life cycle assessment for material selection: A case study of building materials. *International Journal Life Cycle Assessment, 19*, 620–645.

16. Siebert, A., Bezama, A., O'Keeffe, S., & Thrän, D. (2018). Social life cycle assessment: In pursuit of a framework for assessing wood-based products from bioeconomy regions in Germany. *The International Journal of Life Cycle Assessment, 23*(3), 651–662.

17. Mancini, L., Eynard, U., Eisfeldt, F., Ciroth, A., Blengini, G., & Pennington, D. (2018). *Social assessment of raw materials supply chains. A life-cycle-based analysis*, JRC Technical report. Luxembourg: Publications Office of the European Union.

18. Mancini, L., & Sala, S.(2018). Social impact assessment in the mining sector: Review and comparison of indicators frameworks. *Resource Policy.* https://doi.org/10.1016/j.resourpol.2018.02.002.

19. Benoit-Norris, C., Cavan, D. A., & Norris, G. (2012). Identifying social impacts in product supply chains: Overview and application of the social hotspot database. *Sustainability, 4*, 1946–1965. http://www.mdpi.com/2071-1050/4/9/1946/pdf

20. Swami, A., Chopra, V. P., & Maliket, S. L. (2009). Occupational health hazards in stone quarry workers: A multivariate approach. *Journal of Human Ecology, 5*(2), 97–103. https://doi.org/10.1080/09709274.1994.11907078

21. Olusegun, O., Adeniyi, A., & Adeola, G. T. (2009). Impact of granite quarrying on the health of workers and nearby residents in Abeokuta Ogun State, Nigeria. *Ethiopian Journal of Environmental Studies and Management, 2*(1). https://doi.org/10.4314/ejesm.v2i1.43497

22. Nwibo, A. N., Ugwuja, E. I., Nwambeke, N. O., Emelumadu OF, & Ogbonnaya, L. U. (2012). Pulmonary problems among quarry workers of stone crushing industrial site at Umuoghara, Ebonyi State, Nigeria. *International Journal of Occupational Environmental Medicine, 3*(4), 178–185.

23. Ugbogu, O. C., Ohakwe, J., & Foltescu, V. (2009). Occurrence of respiratory and skin problems among manual stone-quarrying workers. *Mera: African Journal of Respiratory Medicine*, 23–26.

24. Ezisi, C. N., Eze, B. I., Okoye, O., & Arinze, O. (2017). Correlates of stone quarry workers' awareness of work-related ocular health hazards and utilization of protective eye devices: Findings in Southeastern Nigeria. *Indian Journal of Occupational Environmental Medicine, 21*(2), 51–55. https://doi.org/10.4103/ijoem.IJOEM_171_16

25. Oginyi, R. C. N. (2010). Occupational health hazards among quarry employees in ebonyi state, Nigeria: Sources and health implications. *International Journal of Development and Management Review (INJODEMAR), 5*(1).

26. Sanmiquel, L., Bascompta, M., Rossell, J. M., Anticoi, H. F., & Guash, E. (2018). Analysis of occupational accidents in underground and surface mining in Spain using data mining techniques. *International Journal of Environmentel Research Public Health, 15*(3), 462. https://doi.org/10.3390/ijerph15030462

Open Access This chapter is licensed under the terms of the Creative Commons Attribution 4.0 International License (http://creativecommons.org/licenses/by/4.0/), which permits use, sharing, adaptation, distribution and reproduction in any medium or format, as long as you give appropriate credit to the original author(s) and the source, provide a link to the Creative Commons license and indicate if changes were made.

The images or other third party material in this chapter are included in the chapter's Creative Commons license, unless indicated otherwise in a credit line to the material. If material is not included in the chapter's Creative Commons license and your intended use is not permitted by statutory regulation or exceeds the permitted use, you will need to obtain permission directly from the copyright holder.

A Life Cycle-Based Scenario Analysis Framework for Municipal Solid Waste Management

Ioan-Robert Istrate, José-Luis Gálvez-Martos, and Javier Dufour

Abstract A framework for the systematic analysis of the material flows and the life cycle environmental performance of municipal solid waste (MSW) management scenarios is described in this article. This framework is capable of predicting the response of waste treatment processes to the changes in waste streams composition that inevitably arise in MSW management systems. The fundamental idea is that the inputs (raw materials and energy) and outputs (final products, emissions, etc.) into/from treatment processes are previously allocated to the specific waste materials contained in the input waste stream. Aggregated indicators like life cycle environmental impacts can then be allocated to waste materials, allowing systematic scenario analyses. The framework is generic and flexible, and can easily be adapted to other types of assessments, such as economic analysis and optimization.

1 Introduction

Municipal solid waste (MSW) is generally defined as that generated in households and from commercial, institutional, and street cleaning activities with similar composition to the household waste. MSW contains a wide variety of potentially valuable materials (e.g., food waste, paper, cardboard, plastic, and metals) but whose increased generation and inappropriate management cause negative environmental and human health consequences as well as the loss of resources [1]. Decision-makers are under increasing pressure to adopt MSW management strategies aiming to maximize resource and energy recovery and minimize environmental and human health risks and usually under constrained budget. In Europe, the implementation of

I.-R. Istrate (✉) · J. Dufour
IMDEA Energy, Systems Analysis Unit, Móstoles, Spain

Rey Juan Carlos University, Chemical and Environmental Engineering Group, Móstoles, Spain
e-mail: robert.istrate@imdea.org

J.-L. Gálvez-Martos
IMDEA Energy, Systems Analysis Unit, Móstoles, Spain

© The Author(s) 2022
Z. S. Klos et al. (eds.), *Towards a Sustainable Future - Life Cycle Management*,
https://doi.org/10.1007/978-3-030-77127-0_20

waste strategies to meet MSW targets is mandatory for the Member States. Reuse and recycling of MSW shall reach 65% by 2035 (Directive 2018/851/EC), while the Circular Economy Package refers to a maximum of 10% landfilling by 2035. However, only 30% of the MSW generated in 2017 in Europe was recycled while the average landfill rate was 23%, even though half of the Member States landfilled more than 50% of their MSW [2].

Handling the complexity of the MSW management system, which encompasses a large number of interconnected processes, remains the main challenge to the development of sustainable MSW management strategies. The waste streams derived from MSW collection and the intermediate waste streams present a hetero-geneous composition of a wide variety of waste materials with different physico-chemical and biological properties (Fig. 1). The resource and energy recovery rates and the technical, economic, and environmental performance of treatment processes depend to a large extent on the composition and properties of the input waste stream [3]. For example, the global warming impact of landfilling the residual waste stream depends on its content on biodegradable waste materials (food waste, paper, etc.), whereas the global warming impact of its incineration depends on its content on fossil-based waste materials (plastic).

Systems analysis techniques are required to tackle the complexity of the MSW management system and support the design of sustainable waste strategies [4]. Life cycle assessment (LCA) has emerged as the most popular, and there are a number of waste LCA tools available. The ability of linking the life cycle inventory (LCI) of treatment processes (i.e., emissions and resource consumption/recovery) to the composition and properties of the input waste stream was recognized as the key feature of a waste-specific LCA tool [5]. However, most of these tools have been developed with black-box models of treatment processes where inputs and outputs are only linked by unrealistic ratios to total mass of input waste. Recently, increased attention is being devoted to the development of modeling frameworks that allow linking input waste composition, treatment process operation, and outputs through a more appropriate approach to physicochemical and biological mechanistic models [6]. This is achieved by adopting a material flow analysis (MFA) perspective for the modeling of the LCIs of treatment processes [7].

MFA is the central methodology of the industrial ecology, and its goal is to pro-vide a comprehensive and systematic inventory of the input-output flows of

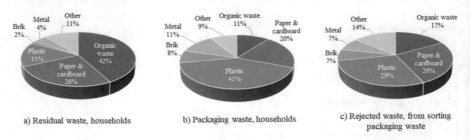

a) Residual waste, households   b) Packaging waste, households   c) Rejected waste, from sorting packaging waste

Fig. 1 Composition of municipal solid waste streams generated in Madrid (2017)

materials and substances in a system. Mass conservation is the fundamental principle of MFA, i.e., the quantity of input flows has to be equal to the quantity of output flows plus stocks [8]. Thus, MFA provides the appropriate mathematical relationships that describe the mass balance of waste materials and their chemical elements in a specific treatment process as well as the parameters required.

In addition to LCA and MFA, numerous optimization models for MSW management have also been developed. Mathematical programming techniques can provide a powerful framework that considers all the feasible configurations of the MSW management system and identify the best solution according to one or multiple objectives and considering the system's constraints. Typically, optimization models focused on economic objectives, e.g., the minimization of the system's annual cost. Also, the additional consideration of environmental objectives (based on LCA) and resource recovery objectives (based on MFA) has emerged as a recent trend [9]. In order to provide reliable results, optimization models should, as in the case of LCA tools, be able to capture the response of treatment processes to changes in the composition of the input waste stream [10, 11]. However, the incorporation of this feature leads to complex nonlinear optimization models, and therefore this issue remains little explored so far.

In this article, we describe a framework for the systematic analysis of the material flows and the life cycle environmental performance of MSW management scenarios. The framework is capable of predicting the response of treatment processes to the changes in waste composition that inevitably arise in MSW management systems. Furthermore, the framework is sufficiently generic and flexible to allow incorporating other methods into the assessment, such as economic analysis and optimization. Section 2 describes the framework. Section 3 includes an illustrative example of its application. Section 4 draws the main conclusions and the future work.

2 Framework Description

2.1 Scope and System Boundaries

Figure 1 illustrates the scope and system boundaries of the framework. Based on the definition of MSW given in the Waste Framework Directive, we considered the waste generated by three sectors: households, commercial activities, and street cleaning. Waste collection at each sector can be defined by combining the five waste streams that could be found in Spanish municipalities: residual, packaging, paper and cardboard, glass, and organic wastes (Fig. 2a). These streams need to be defined in terms of quantity and composition. Waste streams composition is disaggregated into 15 materials: food waste, green waste, mix paper, cardboard, polyethylene terephthalate (PET), high-density polyethylene (HDPE), low-density polyethylene (LDPE), mix plastic, cartons and alike, glass, ferrous metal, nonferrous metal, textile, wood, and other. Furthermore, each waste material is characterized by 83

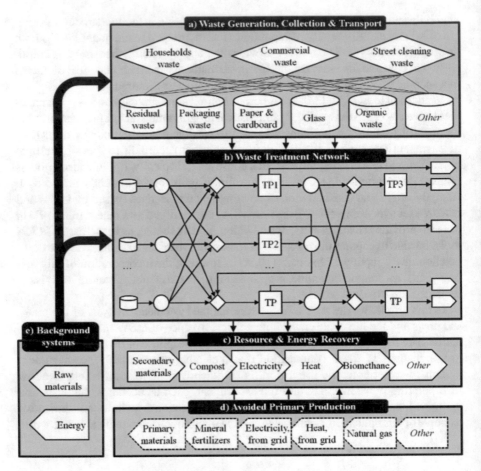

Fig. 2 Scope and system boundaries of the framework

physicochemical and biological properties (e.g., moisture content, lower heating value, biochemical methane potential, chemical elements, etc.).

Collected waste streams are processed in a network of interconnected treatment processes (material recovery facilities, composting, incineration, landfill, etc.) that generate intermediate waste streams (rejected waste, recyclable materials, etc.) and/ or final products (secondary materials, compost, electricity, etc.) (Fig. 2b). While intermediate waste streams need further processing, final products are introduced into the market, thus avoiding primary production (Fig. 2c–d). Additionally, the MSW management system interacts with the background systems that supply raw materials and energy (Fig. 2e). Further details about network structure are provided in Sect. 2.3.

2.2 Modular Modeling

We adopted a modular approach so that the MSW management system was disaggregated into many modules that describe treatment processes [12]. This approach has the advantage that allows combining many technological alternatives. For example, anaerobic digestion (AD) was disaggregated into one module that includes the pre-treatment, reactor, dewatering, and post-treatment unit processes and other four modules for each unit process for the use of the biogas (flare, boiler, combined heat and power, and upgrading). Thus, AD can be combined with any alternative for biogas utilization.

Modules consist of the mathematical equations that describe mass and energy balances as a function of the properties of the input waste stream and the process operation conditions. The inputs (raw materials and energy for operation) and outputs (intermediate waste streams, final products, emissions, etc.) are allocated to the specific waste materials contained in the input waste stream (Fig. 3a), as explained below.

In LCA terminology, modules aim at performing a multi-input allocation of the LCI between the waste materials contained in the input waste stream. According to the ISO 14040/14044 standards recommendations, the allocation of process inputs and outputs should be based on natural causal relationships. We follow the MFA principles to perform the allocation. For example, transfer coefficients are used to model the transfer of input waste materials into the rejected waste stream and the recyclable materials stream in a sorting process. Emissions are allocated based on the physicochemical and biological properties of the waste material. For example, biogenic CO_2 emissions from waste materials incineration are linked to their biogenic carbon content. Electricity production is calculated for each waste material based on its lower heating value and the process electricity conversion efficiency. For those environmental exchanges where there is no obvious mathematical relationships, allocation is done on a mass basis.

Once allocated the inputs and outputs, aggregated indicators, such as life cycle environmental impacts (i.e., global warming, human toxicity, etc.) or the economic costs (i.e., operation costs, revenues, etc.), can also be allocated to each specific waste material. Therefore, instead of calculating the global warming impact associated with the incineration of 1 tonne of residual waste with a fixed composition, the module calculates the global warming impact associated with the incineration of 1 tonne of each waste material that may constitute the residual waste. Allocated inputs, outputs, and indicators are stored in non-square matrixes that represent in rows the 15 waste materials considered and in columns the inputs, outputs, and indicators given per tonne of waste material (Fig. 3b).

This approach has the advantage that translates the complex nonlinear mathematical models that describe mass and energy balances in treatment processes (e.g., methane generation in landfill is given by a time-dependent first-order decay equation) into a parametrized model (i.e., linear) that can be used for scenario analysis or optimization (Fig. 3c).

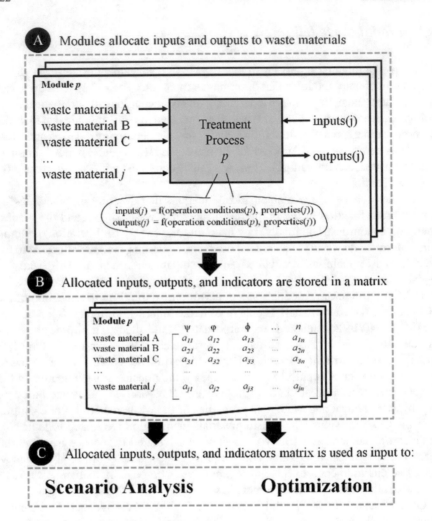

Fig. 3 Modular modeling of treatment processes

2.3 Scenario Analysis Modeling

The modules' matrixes of inputs, outputs, and indicators can be used for scenario analysis. All feasible modules combinations for treating household, commercial, and street cleaning waste streams as well as the intermediate waste streams are embedded in a mathematical network (Fig. 2b). The network consists of splitters (circles), mixers (diamonds), modules (boxes), and all their interconnections (arrows).

Splitters are located after each waste stream and assign the waste stream to the linked modules. The partitioning of a waste stream in a splitter between the modules linked is represented by user-defined mass fractions. For example, the mass fractions of a splitter for residual waste could be 20% to incineration and 80% to landfilling. Note that the waste streams leaving the splitter have the same composition as the input stream because splitters do not involve transformation. Therefore, the mass fraction introduced is applied equally to all the waste materials contained in the waste stream.

Mixers are located after splitters and prior each module. Since modules can receive several waste streams with different composition, the function of mixers is to sum over materials of the same type contained in different waste streams. Mixers do not require input data. Finally, modules performance, for example, the global warming impact of incinerating 20% of the residual waste, results by multiplying the array of input waste materials by the array of global warming impact contained in the matrix of inputs, outputs, and indicators obtained in Sect. 2.2. The performance of the overall MSW management system is obtained by addition of the performance of all modules. Note that, once the allocated inputs, outputs, and indicators for all modules are obtained, the only requirement to build a scenario is to introduce the mass and composition of the initial waste streams and to fill the mass fractions of all splitters in the network.

3 Illustrative Scenario Analysis Case Study

In order to illustrate the applicability of the framework, a streamlined example addressing the global warming consequences of MSW incineration phasing out in Madrid (Spain) is presented. In 2017, about 313,697 t of rejected waste from sorting residual and packaging waste stream at material recovery facilities have been incinerated in Madrid [13]. The new waste strategy of the city aims at phasing out the incineration plant by 2025, which can led to the diversion of huge amounts of waste toward landfilling. In this example, we assess the life cycle global warming impact associated with the management of 1 tonne of rejected waste in Madrid considering different incineration rates. Four scenarios were formulated. S1 considers that 100% of the rejected waste is incinerated. S2 considers that 75% is incinerated and 25% landfilled. S3 considers that 50% is incinerated and 50% landfilled. Finally, S4 considers that 100% is landfilled. The ILCD-recommended characterization factors were used for the assessment [14]. Emissions of biogenic CO_2 and the biogenic carbon that remains sequestered in landfill after 100 years were assumed with a characterization factor of 0.

Table 1 shows the life cycle global warming impact allocated to waste materials as obtained from the incineration and landfilling modules. Incineration was disaggregated into emissions to air (INC [UP_1]), resource consumption (INC [UP_2]), and avoided impacts due to the substitution of electricity from the Spanish mix (INC [UP_3]). Landfilling was disaggregated into dispersive emissions (LAND [UP_1]), resource consumption (LAND [UP_2]), and avoided impacts due to the substitution of electricity from the Spanish mix (LAND [UP_3]). Note that values in Table 1 were computed using technology and operation conditions from Madrid.

Table 1 Life cycle global warming impact allocated to waste materials for incineration (INC) and landfilling (LAND) for the case study of Madrid (kg $CO_{2\text{-eq}}$ t^{-1} wet waste material)

Waste material	INC [UP_1]	INC [UP_2]	INC [UP_3]	LAND [UP_1]	LAND [UP_2]	LAND [UP_3]
Food waste	24	12	−76	733	0.07	−110
Green waste	21	6	−48	301	0.07	−43
Mix paper	11	5	−143	1615	0.07	−149
Cardboard	9	4	−111	1045	0.07	−72
PET	2326	6	−307	0	0.07	0
HDPE	2499	7	−417	0	0.07	0
LDPE	1448	8	−218	0	0.07	0
Mix plastic	2692	22	−417	0	0.07	0
Cartons and alike	140	6	−128	810	0.07	−56
Glass	0	0	0	0	0.07	0
Ferrous metal	0	0	0	0	0.07	0
Nonferrous metal	0	0	0	0	0.07	0
Textile	440	46	−223	240	0.07	−22
Wood	38	9	−206	71	0.07	−4
Other	217	8	−30	0	0.07	0

Table 1 reveals the large differences that exist with respect to the environmental impacts of waste materials. The global warming impact of incinerating plastic is largely higher than other waste materials due to the higher content on fossil carbon. Avoided impacts due to electricity substitution are also higher for plastic due to the higher energy content. Dispersive greenhouse gas emissions from landfill are significantly higher for mix paper, cardboard, and cartons and alike compared to food and green waste. Note that values in Table 1 are expressed per tonne of wet waste material. Although food and green waste have a higher degradation rate compared to paper and cardboard, the former have also higher moisture content. Finally, the global warming impact of resource consumption in landfill (electricity and diesel for landfill operation) is the same for all waste materials. This reflects that energy consumption was allocated on a mass basis because energy is used for waste movement. Consequently, the same impact is obtained per tonne of each waste material.

Figure 4 shows the procedure to build scenarios S1–S4 and how the global warming impact of each scenario is calculated. Quantity (Q) and composition (c) are required as input data in order to define the rejected waste stream. The quantity was assumed 1 wet tonne (functional unit), and the composition is as follows: 13.91% food waste, 3.52% green waste, 26.48% mix paper, 8.46% cardboard, 2.54% PET, 1.30% HDPE, 10% LDPE, 7.65% mix plastic, 3.71% cartons and alike, 3.57% glass, 1.49% ferrous metal, 1.11% nonferrous metal, 10.98% textile, 5.29% wood, and 0% other. The input data into the splitter are the mass fractions of the rejected waste stream to incineration (σ) and landfilling (ω). The allocated global warming impact of incineration INC(GW) and landfilling LAND(GW) were calculated by

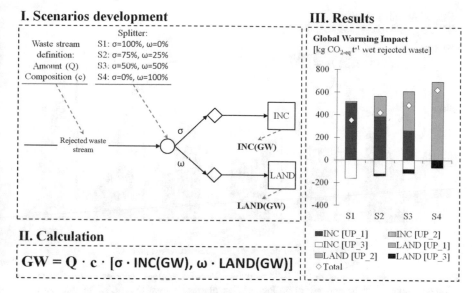

Fig. 4 Scenarios development (I), calculation (I), and global warming impact results (III) for the scenarios addressed

the framework (Table 1). Thus, the global warming impact of each scenario disaggregated by unit processes can be easily obtained.

For this case study, increasing the landfilling of rejected waste at the expense of reducing incineration entails an increase in the global warming impact. The increase is related to the dispersive emissions of methane from landfill. Note that the mix paper and cardboard contained in the rejected waste are significant: 26.48% and 8.46%, respectively. These waste materials have the highest global warming impact on landfilling. In contrast, their impact on incineration is negligible because biogenic CO_2 emissions were considered not to contribute to the global warming impact (Table 1).

Figure 5 shows the global warming impact of S1–S4 as a function of a gradual decrease on mix paper and cardboard content at the expense of an increase on plastic content. The results highlight the key role of waste composition when assessing MSW management systems. In fact, if the rejected waste did not contain mix paper and cardboard but a higher proportion of plastic, landfilling would be a better option than incineration. This is because the global warming impact of plastic in landfill is negligible (Table 1).

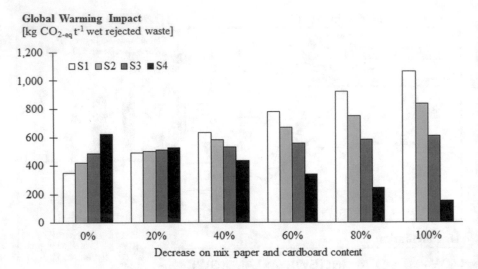

Fig. 5 Global warming impact of S1–S4 as a function of a gradual decrease on mix paper and cardboard content at the expense of an increase on plastic content. (S1, 100% incineration; S2, 75% incineration and 25% landfill; S3, 50% incineration and 50% landfill; S4, 100% landfill)

4 Conclusions and Future Work

A framework for the systematic analysis of the material flows and the life cycle environmental performance of municipal solid waste (MSW) management scenarios has been proposed and described in this article. The framework addresses the collection, treatment, and final disposal of household, commercial, and street cleaning waste streams generated in a given region. System boundaries include the network of interconnected treatment processes, the recovery of resource and energy that avoid primary production, as well as other background systems that supply raw materials and energy to the MSW management system.

The framework is based on a modular modeling approach so that the MSW management system was disaggregated into many modules that describe treatment processes (or even stages of treatment processes). All feasible modules combinations are embedded in a network, and therefore any (feasible) MSW management scenario can be addressed. A key feature of the framework is its capability of tackling the assessment of the complex response of treatment processes to the changes in waste streams composition that inevitably arise in MSW management. The fundamental idea is that inputs (raw materials and energy for operation), outputs (final products, emissions, etc.), and aggregated indicators (life cycle environmental impacts, economic costs, etc.) of treatment processes are previously allocated to the specific waste materials contained in the input waste stream.

The framework is generic and flexible to the incorporation of other types of assessments. The allocated inputs, outputs, and indicators can be used as input parameters into an optimization model. This represents an enormous advantage since the response of treatment processes to changes in waste composition can be easily evaluated with fixed parameters. There is no need to formulate a mathematical program based on the complex nonlinear models that describe mass and energy balances in waste treatments. The only requirement is to consider as optimization variables the flow of each waste material contained in waste streams. While the modeling approach based on the flow of multi-components has been typically applied in wastewater networks optimization problems, this remains unexplored in the field of MSW management.

Acknowledgments This research has been supported by the Spanish Ministry of Science, Innovation and Universities (RTI2018-097227-B-I00) and a project co-financed by the Connecting Europe Facility of the European Union (Grant Agreement number INEA/CEF/TRAN/M2016/1359344).

References

1. Kaza, S., Yao, L., Bhada-Tata, P., & Van Woerden, F. (2018). *What a waste 2.0: A global snapshot of solid waste management to 2050, urban development series*. World Bank.
2. Eurostat, Municipal waste statistics (Accessed 10.01.2020).
3. Bisinella, V., Götze, R., Conradsen, K., Damgaard, A., Christensen, T. H., & Astrup, T. F. (2017). Importance of waste composition for Life Cycle Assessment of waste management solutions. *Journal of Cleaner Production, 164*, 1180–1191.
4. Pires, A., Martinho, G., & Bin Chang, N. (2011). Solid waste management in European countries: A review of systems analysis techniques. *Journal of Environmental Management, 92*(4), 1033–1050.
5. Gentil, E. C., Damgaard, A., Hauschild, M., Finnveden, G., Eriksson, O., Thorneloe, S., Ozge Kaplan, P., Barlaz, M., Muller, O., Matsui, Y., Ii, R., & Christensen, T. H. (2010). Models for waste life cycle assessment: Review of technical assumptions. *Waste Management, 30*(12), 2636–2648.
6. Lodato, C., Tonini, D., Damgaard, A., & Astrup, T. F. (2020). A process-oriented life-cycle assessment (LCA) model for environmental and resource-related technologies (EASETECH). *International Journal of Life Cycle Assessment, 25*(1), 73–88.
7. Turner, D. A., Williams, I. D., & Kemp, S. (2016). Combined material flow analysis and life cycle assessment as a support tool for solid waste management decision making. *Journal of Cleaner Production, 129*, 234–248.
8. Brunner, P. H., & Rechberger, H. (2004). *Practical handbook of material flow analysis*. Lewis Publisher.
9. Juul, N., Münster, M., Ravn, H., & Ljunggren Söderman, M. (2015). Economic and environmental optimization of waste treatment. *Waste Management, 38*(1), 486–495.
10. Levis, J. W., Barlaz, M. A., DeCarolis, J. F., & Ranjithan, S. R. (2013). A generalized multistage optimization modeling framework for life cycle assessment-based integrated solid waste management. *Environmental Modelling & Software, 50*, 51–65.
11. Roberts, K. P., Turner, D. A., Coello, J., Stringfellow, A. M., Bello, I. A., Powrie, W., & Watson, G. V. R. (2018). SWIMS: A dynamic life cycle-based optimisation and decision support tool for solid waste management. *Journal of Cleaner Production, 196*, 547–563.

12. Haupt, M., Kägi, T., & Hellweg, S. (2018). Modular life cycle assessment of municipal solid waste management. *Waste Management, 79*, 1–13.
13. Madrid City Council, Memoria de Actividades de la Dirección General del Parque Tecnológico de Valdemingómez – 2017, 2019.
14. Fazio, S., Castellani, S., Sala, V., Schau, S., Secchi, E., & Zampori, M. (2018). *Supporting information to the characterisation factors of recommended EF Life Cycle Impact Assessment method*. European Commission.

Open Access This chapter is licensed under the terms of the Creative Commons Attribution 4.0 International License (http://creativecommons.org/licenses/by/4.0/), which permits use, sharing, adaptation, distribution and reproduction in any medium or format, as long as you give appropriate credit to the original author(s) and the source, provide a link to the Creative Commons license and indicate if changes were made.

The images or other third party material in this chapter are included in the chapter's Creative Commons license, unless indicated otherwise in a credit line to the material. If material is not included in the chapter's Creative Commons license and your intended use is not permitted by statutory regulation or exceeds the permitted use, you will need to obtain permission directly from the copyright holder.

The Life Cycle Sustainability Indicators for Electricity Generation in Chile: Challenges in the Use of Primary Information

Mabel Vega-Coloma and Claudio Zaror

Abstract The need to get an appropriate quantification of the sustainability indicators involves the use of site-specific information that could come from several sources, affecting its quality. This study analyses the quality and sources to build eight environmental, seven social and four economic indicators for eight electricity generation technologies in 2005, 2009 and 2015 as reference years, following the ISO 14.040-44:2006 life cycle assessment approach. The results show for the three dimensions important differences among the periods, reaching over 400% of reduction in 2015 in case of acidification for coal power plants, thanks to environmental regulations. For levelized electricity cost and corruption index, the variations reach around 40% and 30%, mainly for fossil fuel-based power plants. These changes support the need to have a centralized, reliable and accurate data system of registration, in order to contribute to the sustainability of the electricity system in Chile.

1 Introduction

The need to get an appropriate quantification of the sustainability indicators involves the use of site-specific information [1]. This information could come from several sources and sometimes is barely systematized and highly heterogonous, being its quality and consistency a matter of concern [2]. Due to the increasing environmental, economic and social requirements, more data are available to model the potential impacts profile. In particular, the power plants of electricity generation in Chile report continuously their air and water emissions, as well as the hazardous waste

M. Vega-Coloma (✉)
Escuela de Ingeniería Química, Facultad de Ingeniería, Universidad del Bío-Bío,
Concepción, Chile
e-mail: mvega@ubiobio.cl

C. Zaror
Departamento de Ingeniería Química, Facultad de Ingeniería, Universidad de Concepción,
Concepción, Chile

© The Author(s) 2022
Z. S. Klos et al. (eds.), *Towards a Sustainable Future - Life Cycle Management*,
https://doi.org/10.1007/978-3-030-77127-0_21

generated from their process, and from those were developed several accurate reports assessing the environmental performance [3–6]. Nevertheless, the results could be a source of more questions about the data quality, such as the traceability or the methodological approach to measure mass fluxes [1, 2, 7]. In the same way, the economic profile is well-known data for experts and investors, but is not always open and available for researchers. The social indicators are still under development and the data is usually scattered. For these reasons, just a few studies had covered jointly the environmental, economic and social dimensions [8–12].

As was reported by Laurent and Espinosa [13] is relevant to study the variability of the environmental performance at the country level, exploring the opportunity to analyse the annual environmental, economic and social performance, just in case to have enough reliable information. This analysis could bring information about the data sources, the quality needed and the potential effect of the assumptions. Moreover, in the case of developing countries, this analysis could be helpful to policy-makers to evidence the legal and regulatory needs to improve the current report of projects.

2 Goal and Scope

The aim of this work is to contribute to the discussion about the use of primary information reported directly from electricity generation power plants, to get an appropriate pool of environmental, economic and social indicators for the electricity generation in Chile. The scope of this work is cradle to gate for the environmental aspects, while for social and economic are covered the direct processes, due to the lack of information.

For environmental issues was included the whole electricity generation process, from the materials and fuel extraction from natural sources to the decommission stage, including the transport, infrastructure and operation specific for each year assessed (see Fig. 1). Economic indicators were developed based on local information for technologies investments, for each year, and reported as "blackbox", without the possibility of disaggregating by stage. In the same sense, the social

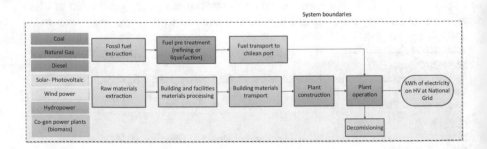

Fig. 1 System boundaries of the electricity generation in Chile considered in this work

information was obtained from several sources and represents different stages of the process (e.g. employment is associated with infrastructure and operation, while import dependency is associated with the whole process).

3 Methodological Procedure

This work was developed following the ISO 14.040-44:2006 [14, 15] approach for life cycle assessment. The electricity generation power plants covered in this work were selected to build a set of eight environmental, four economic and seven social indicators, following a life cycle approach applied to eight electricity generation technologies, coal, diesel, natural gas, biomass, wind power, solar photovoltaic (PV), run of river and reservoir, in Chile. The temporal coverage includes 10 years, using specific data for 2005, 2009 and 2015. The technologies assessed cover more than 99.5% of the current installed capacity in Chile, and the geographical coverage only includes the continental territory, excluding Patagonia. The electricity generation produced by technology for the period analysed is presented in Table 1.

3.1 Definition of Environmental, Economic and Social Indicators

The environmental indicators have been calculated from CML 2000 mid-point impact model. Several studies have worked with this impact model to represent the damage over different categories. Some categories are associated with environmental impacts and another with social impacts, as is detailed in respective subsection.

Eight environmental indicators, namely, ozone layer depletion potential (ODP), photochemical oxidation potential (POP), global warming potential (GWP), acidification potential (AP), eutrophication potential (EP), freshwater aquatic ecotoxicity

Table 1 Electricity generation in Chile in 2005, 2009 and 2015

| Technology | Electricity generation by source (GWh/y) | | |
	2005	2009	2015
Coal	8813	15,625	28,613
Diesel	1113	1395	2862
Natural gas	14,681	1,444,038	10,807
Biomass	518	968	1931
Wind power	–	61	2103
Solar PV	–	–	1373
Reservoir	14,801	13,921	11,616
Run of river	10,673	10,633	12,283
Total	50,599	56,641	71,588

potential (FAEP), marine aquatic ecotoxicity potential (MAETP) and terrestrial eco-toxicity potential (TEP) impacts were assessed in this study, on the basis of previous work [3, 4]. These indicators were calculated following the ISO 14.040-44:2006 standards for life cycle assessment [14, 15], using the CML 2000 v.2.05 mid-point impact models [16], with the computational support of SimaPro v.7.3.3 software [17].

On the other hand, four indicators were used to address economic issues, namely, total capital cost (TCC), levelized electricity cost (LEC) and fuel sensitivity price (FSP) as proposed by [18], while total annualized cost (TAC) was considered from the definition brought by [9]. Finally, seven indicators related to social issues were estimated. These issues were addressed by [18] and categorized as follows:

- Energy security, measured as import dependency (ID), imported fuels potentially avoided (IFPA) and diversity of fuel supply (DFS)
- Provision of employment (PE)
- Intergenerational issues, measured as human toxicity (HT) and abiotic deple-tion (ADP)
- Local community impacts measured as corruption index (CI)

Every indicator was estimated by technology and by year totalizing 399 indica-tors specific for Chilean electricity situation.

3.2 Data Sources, Quality and Assumptions

The data were obtained mainly from primary open sources. They were several gov-ernmental offices and institutions, which have been implemented a transparency system of data registration, mainly driven by environmental control regulations. In this sense, was possible to get reliable data from these sources to build the most part the indicators reported [19–24]. Some others were obtained from studies [25], inter-national reports [26] and assumptions.

The most part of the assumptions were addressed to economic and social indica-tors. Particularly, all the costs were corrected to 2015 value, considering the infla-tion, in order to compare the decade's values. Due the lack of data for cost of renewables investment in Chile, they were assumed from international values for the same technology [26]. For corruption index, the data were considered using the perception index for the respective year, and the mix of imported fuels.

3.3 Variation on Indicator Values

For every indicator, the variation with respect to the 2015 value was estimated, in order to represent a better or worst situation in the past compared with the current. To quantify this, the use of a percentage of variation is proposed defined by Equation 1.

$$\%variation = \left(\frac{Indic.value_{2005} - Indic.value_{2015}}{Indic.value_{2015}} \right) *100 \qquad (1)$$

This percentage represents the pathway that every single indicator has followed during this last decade. Depending of the accuracy and representability of the information reported for each indicator, this variation could be relevant or null.

This variation does not apply to solar PV during 2005 and 2009, as well as for wind power during 2005 due the lack of contribution to the electricity generation from these sources in those years.

4   Results and Discussion

The results are presented in terms of a general analysis of the data quality followed by the main variation of the environmental, economic and social indicators among the period covered between 2005 and 2015.

4.1   Analysis of Data Quality and Sources

The analysis of the data quality is based on the description of the five aspects included in the pedigree matrix. The description is presented in Table 2.

From the table above, it is possible to identify that only reliability could be a source of uncertainty, while the rest are well covered. However, the data

Table 2   Description of data quality based on pedigree matrix aspects

Pedigree aspects	Description
Reliability	The most part of environmental, economic and social data were obtained from primary source, with exception of infrastructure of run of river, diesel and solar PV, which were complemented with ecoinvent data. For the economic indicator, only the costs for biomass were obtained from foreign source in 2005 and 2009
Completeness	All the process stages were covered in the simulation and there are no missing data
Temporal coverage	Each environmental, economic and social data was specific for each year and not average was used covering more than 1 year
Geographical coverage	The continental territory in Chile was completely covered with the exception of Patagonia. The overseas territory was not included
Technological coverage	The 99.5% of installed capacity of each year was covered in this study, upgrading annually the electricity generation, the conversion efficiency, air emissions, prices and corruption and perception index

assumptions are based on verified information, becoming the main constrains the source and its location instead of the availability of data.

In this case, the location of data process for the environmental profile by technology was based on sources from the Environmental and Energy Ministry and official agencies. This information must be reported because it is mandated by law. On the contrary, the information for the economic profile were obtained from different and heterogeneous sources, such as official agencies, international reports and specific studies, avoiding the systematization of the consultancy and the respective updating. The Economic Ministry has no data about these issues and the Energy Ministry has only information about the cost of investment just for some years and not enough disaggregated to be consistent with the life cycle approach. In fact, this information could be really well-known by investors but still unknown by the researchers. More dramatic is the availability of the social information, with no official institution or agency controlling these aspects and main part of the information obtained from sources related with environmental issues. Currently, scatter and spot information are available determining the capacity to include more and better social indicators.

Everything was possible only thanks to a very detailed and continuously updated knowledge about current instruments of central information report, which has been very dynamic and improved. For this reason, the development of a detailed work like this could be a field of close relation with the central authorities, in order to have a constant validation and evidence the future needs.

4.2 Environmental Indicators

The environmental indicators had important changes during the period, mainly due to changes for conversion efficiency by technology and changes in the environmental regulations. The effect of the conversion efficiency is shown in Fig. 2, for natural gas mainly, where in 2005 and 2009 the acidification potential (ACP) was 10% and 22% higher than in 2015, respectively. On the other hand, the effect of new environmental regulations over the environmental profile of the thermal technologies is clearly exposed in the behaviour of acidification for coal power plants in 2005 and 2015. The reduction of this indicator was over 400%, thanks to a specific regulation for thermoelectric plants, due to its high emission levels and poor abatement systems.

The rest of the environmental indicators present changes like that representing the specific annual situation of each technology [3]. Since these relevant changes in each indicators, it worth to keep constantly evaluated the process data reported in the environmental system, in order to validate them and contribute with a more accurate and transparent central information repository.

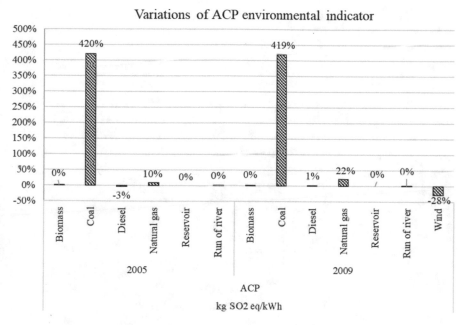

Fig. 2 Variation percentage of acidification environmental indicator in 2005 and 2009 in relation to 2015

4.3 Economic Indicators

Economic indicators are very sensitive with market constrains; for this reason, the need to count with updated information is vital. As is possible to see from Fig. 3, the levelized electricity cost presents a relevant variation in 2005 and 2009 relative to 2015 for all technologies. In fact, thermal power plants present higher values in 2015 than in 2005 and 2009, reaching for coal −36% and −30%, respectively. Due to the uncertainty related to the cost of investment for biomass, the fluctuations are very wide. The reduction of the LEC of wind power among 2009 and 2015 presents the trends in the international markets, where the cost of this technology has decreased.

The important changes in the economic indicators reflect the need to have an updated source of information based on local restrictions, where it could be systematized in order to contribute to an accurate economic analysis. The use of average values from other countries are too vague, for the same reason that is not convenient to consider the economic allocation for the environmental burdens [14, 15].

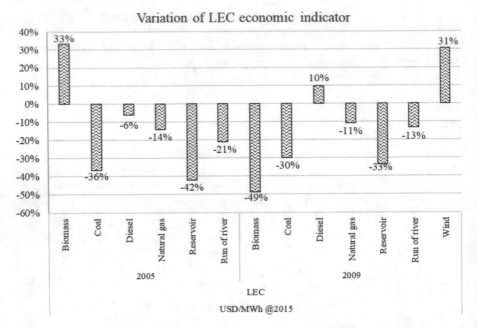

Variation of LEC economic indicator

Fig. 3 Variation percentage of levelized electricity cost economic indicator in 2005 and 2009 in relation to 2015

4.4 Social Indicators

The social indicators present an important variation among the period. In Fig. 4 is shown the corruption index (CI) performance, which is very sensitive for coal, natural gas, diesel and biomass, mainly due to the importance of fuel. In the case of biomass, changes in corruption and perception index in 2005 and 2009 explain the different trends through those years. In the same sense, coal presents the same different trends. While in 2005 coal was imported from Australia and Canada mainly, in 2009, there was a change as to the importing country which was exclusively done by Colombia, to be finally shared in 2015 with USA and Australia.

In the same sense as environmental and economic indicators, social indicators present a dynamic behaviour which could be dramatically different when the conditions of technologies present changes through time.

These evidences are key to sustain the need to have a data depository which can be systematized, appropriate, reliable and accurate, in order to take advantage of the current process data report, to shift them to a sustainable platform with updated and continuously improved data accuracy and to assess the global performance, specifically, of the electricity sector in Chile.

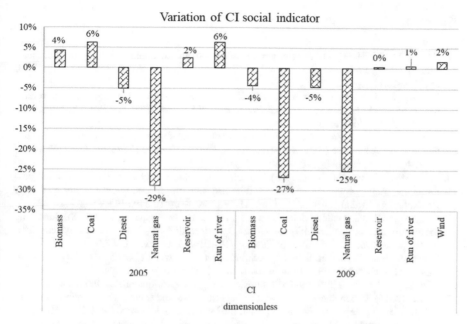

Fig. 4 Variation percentage of corruption index social indicator in 2005 and 2009 in relation to 2015

5 Conclusions

The knowledge about the local environmental reporting system could be really helpful in the case of build a database with primary data regarding the life cycle approach for processes. This system for Chile is still a matter of development, but results like these could be an important input in order to address improvements to the current system. Nevertheless, the social and economic issues are still scattered and were obtained from heterogeneous sources, which are not necessarily the best option for a detailed assessment. This is closely related with the legal need to report the operational performance instead of the global performance.

Understanding sustainability as an equilibrium between environmental, social and economic dimensions, the development of the electric sector ought to be driven to improve data quality, systematizing the reports to check and manage the global performance of the sector.

This critical analysis could be useful for decision-makers and countries in the pathway of development, which are implementing environmental open reports with perational data, specifically for the electricity generation.

References

1. Curran, M. A., Mann, M., & Norris, G. (2005). The international workshop on electricity data for life cycle inventories. *Journal of Cleaner Production, 13*(8), 853–862. https://doi.org/10.1016/j.jclepro.2002.03.001
2. Hellweg, S., & Milà i Canals, L. (2014). Emerging approaches, challenges and opportunities in life cycle assessment. *Science (New York, N.Y.), 344*(6188), 1109–1113. https://doi.org/10.1126/science.1248361
3. Vega-Coloma, M., & Zaror, C. A. (2018a). Environmental impact profile of electricity generation in Chile: A baseline study over two decades. *Renewable and Sustainable Energy Reviews, 94*, 154–167. https://doi.org/10.1016/j.rser.2018.05.058
4. Vega, M. I., & Zaror, C. A. (2018b). The effect of solar energy on the environmental profile of electricity generation in Chile: A midterm scenario. *International Journal of Energy Production and Management, 3*(2), 110–121. https://doi.org/10.2495/EQ-V3-N2-110-121
5. Gaete-Morales, C., Gallego-Schmid, A., Stamford, L., & Azapagic, A. (2018). Assessing the environmental sustainability of electricity generation in Chile. *Science of the Total Environment, 636*, 1155–1170. https://doi.org/10.1016/j.scitotenv.2018.04.346
6. Gaete-Morales, C., Gallego-Schmid, A., Stamford, L., & Azapagic, A. (2019). Life cycle environmental impacts of electricity from fossil fuels in Chile over a ten-year period. *Journal of Cleaner Production, 232*, 1499–1512. https://doi.org/10.1016/j.jclepro.2019.05.374
7. UNEP. (2016). Green energy choices: The benefits, risks and trade-offs of low-carbon technologies for electricity production. Report of the International Resource Panel. E. G. Hertwich, J. Aloisi de Larderel, A. Arvesen, P. Bayer, J. Bergesen, E. Bouman, T. Gibon, G. Heath, C. Peña, P. Purohit, A. Ramirez, S. Suh, (eds.).
8. Atilgan, B., & Azapagic, A. (2016). An integrated life cycle sustainability assessment of electricity generation in Turkey. *Energy Policy, 93*, 168–186. https://doi.org/10.1016/j.enpol.2016.02.055
9. Santoyo-Castelazo, E., & Azapagic, A. (2014). Sustainability assessment of energy systems: Integrating environmental, economic and social aspects. *Journal of Cleaner Production, 80*, 119–138. https://doi.org/10.1016/j.jclepro.2014.05.061
10. Stamford, L., & Azapagic, A. (2014). Energy for Sustainable Development Life cycle sustainability assessment of UK electricity scenarios to 2070. *Energy for Sustainable Development, 23*, 194–211. https://doi.org/10.1016/j.esd.2014.09.008
11. Maxim, A. (2014). Sustainability assessment of electricity generation technologies using weighted multi-criteria decision analysis. *Energy Policy, 65*, 284–297. https://doi.org/10.1016/j.enpol.2013.09.059
12. Santos, M. J., Ferreira, P., Araújo, M., Portugal-pereira, J., Lucena, A. F. P., & Schaeffer, R. (2017). Scenarios for the future Brazilian power sector based on a multi- criteria assessment. *Journal of Cleaner Production, 167*, 938–950. https://doi.org/10.1016/j.jclepro.2017.03.145
13. Laurent, A., & Espinosa, N. (2015). Environmental impacts of electricity generation at global, regional and national scales in 1980–2011: What can we learn for future energy planning? *Energy Environmental Science, 8*(3), 689–701. https://doi.org/10.1039/C4EE03832K
14. ISO, International Standardization Organization. (2006a). Environmental management – ISO 14.040. Life cycle Assessment – Principles and framework.
15. ISO, International Standardization Organization. (2006b). Environmental management – ISO 14.044. Life cycle Assessment – Requirements and guidelines.
16. Guinée, J. B., Gorrée, M., Heijungs, R., Huppes, G., Kleijn, R., Koning, A. de, Oers, L. van, Wegener Sleeswijk, A., Suh, S., Udo de Haes, H.A., de Bruijn, H., van Duin, R., Huijbregts, M.A.J. (2002) Handbook on life cycle assessment. Operational guide to the ISO standards. I: LCA in perspective. IIa: Guide. IIb: Operational annex. III: Scientific background. Kluwer Academic Publishers, ISBN 1-4020-0228-9, Dordrecht, 692 pp.

17. PRé Sustainability. Eco-Indicator 99, Manual for Designers (2000). Pré Sustainability, Amersfoort, The Netherlands. https://www.pre-sustainability.com/download/EI99_Manual.pdf
18. Stamford, L., & Azapagic, A. (2011). Sustainability indicators for the assessment of nuclear power. *Energy, 36*(10), 6037–6057. https://doi.org/10.1016/j.energy.2011.08.011
19. CNE, Comisión Nacional de Energía, Gobierno de Chile. (2018). Balance nacional de energía año 2017. http://energiaabierta.cl/visualizaciones/balance-de-energia/ (Accessed in January 2020)
20. SEA, Sistema de Evaluación Ambiental, Gobierno de Chile. (2016). Sistema de evaluación de impacto ambiental. http://www.sea.gob.cl/ (Accessed in October- November 2016).
21. Aduanas, Gobierno de Chile. (2016). Registros agregados de comercio exterior. https://www.aduana.cl/registros-de-comercio-exterior-datos-agregados/aduana/2017-07-21/113048.html (Accessed in October- November 2016).
22. RETC, Registro de emisiones y transferencia de contaminantes, Gobierno de Chile. (2016). http://www.retc.cl/datos-retc/ (Accessed in October- december 2016).
23. CDEC-SING, Centro de despacho económico de carga, Sistema Interconectado Norte Grande. (2016). Anuario y estadísticas de operación año 2015. http://cdec2.cdec-sing.cl/html_docs/anuario2015/sing.html (Accessed in August 2015).
24. CDEC-SIC, Centro de despacho económico de carga, Sistema Interconectado Central. (2016). Anuario y estadísticas de operación año 2015. https://sic.coordinador.cl/wp-content/uploads/2016/04/SIC_2015.pdf (Accessed in August 2016).
25. Bennet, M. Pérez, H. (2009). Cambio de la matriz energética chilena en relación a la señal de precios. Departamento de Ingeniería Eléctrica. Pontificia Universidad Católica de Chile. http://hrudnick.sitios.ing.uc.cl/alumno09/matriz/Evolucion%20de%20la%20Matriz%20Energetica.pdf (Accessed 12.06.18)
26. IEA, International Energy Agency. (2016). Projected costs of generating electricity, 2015 Edition. 30749September 2015 edition. https://www.iea.org/Textbase/nptoc/ElecCost2015TOC.pdf750 (Accessed on February and March 2018)

Open Access This chapter is licensed under the terms of the Creative Commons Attribution 4.0 International License (http://creativecommons.org/licenses/by/4.0/), which permits use, sharing, adaptation, distribution and reproduction in any medium or format, as long as you give appropriate credit to the original author(s) and the source, provide a link to the Creative Commons license and indicate if changes were made.

The images or other third party material in this chapter are included in the chapter's Creative Commons license, unless indicated otherwise in a credit line to the material. If material is not included in the chapter's Creative Commons license and your intended use is not permitted by statutory regulation or exceeds the permitted use, you will need to obtain permission directly from the copyright holder.

Translating LCA Evidence into Performance-Based Policy Criteria for the Photovoltaic Product Group

Nieves Espinosa, Nicholas Dodd, and Alejandro Villanueva

Abstract Life cycle assessment has the potential to generate valuable information and knowledge for policy makers, as insights can be gained by applying LCA to the development of policy criteria. This potential has been used in the development of a number of EU policy instruments aimed at photovoltaic products, i.e. Ecodesign, Energy Labelling, the EU Ecolabel and Green Public Procurement. They are the regulatory and voluntary policy instruments for sustainable production and consumption at the European Commission. Each instrument has different market objectives; e.g. Ecodesign sets mandatory minimum requirements for products entering the EU market, while the EU Ecolabel is a voluntary instrument to differentiate the most sustainable choices. An eight-step approach based on the Ecodesign methodology including a systematic LCA review has been used with a focus on the information needs of the policy instruments and an interpretation of the results per component/substance. Through the identification of hotspots at the component level and at life cycle stages, it has been possible to translate them into criteria.

1 Introduction

The EU has a number of legislative instruments which translate EU energy and climate policy goals into various strands of action. Ecodesign and Energy Labelling legislations support the Commission's overarching priority to strengthen Europe's competitiveness and boost job creation and economic growth [1, 2]. They are mandatory instruments that ensure a level playing field in the internal market, drive investment and innovation in a sustainable manner and save money for consumers while reducing CO_2 emissions. These instruments contribute to the Energy Union 2020 and 2030 energy efficiency targets, and to a deeper and fairer internal market.

Two further voluntary policy instruments contribute to fulfil the mentioned objectives: the EU Ecolabel and the Green Public Procurement. The EU Ecolabel (set up under the provisions of Regulation EC 66/2010) aims at reducing the

N. Espinosa (✉) · N. Dodd · A. Villanueva
European Commission, Joint Research Centre, Seville, Spain
e-mail: nieves.espinosa@ec.europa.eu

© The Author(s) 2022
Z. S. Klos et al. (eds.), *Towards a Sustainable Future - Life Cycle Management*,
https://doi.org/10.1007/978-3-030-77127-0_22

negative impact of products and services on the environment, health, climate and natural resources [3]. The EU Ecolabel criteria take into account the environmental improvement potential along the life cycle of products. Green Public Procurement (GPP) is defined in COM(2008)400 as a process whereby public authorities seek to procure goods, services and works with a reduced environmental impact through their life cycle when compared to goods, services and works with the same primary function that would otherwise be procured [4]. GPP takes recently into consideration circular economy aspects in new criteria.

The Ecodesign Working plan that periodically lays out which product groups offer an energy saving potential included in its 2016–2019 edition [5] the photovoltaic group product as one that justified an analysis of the feasibility of potential implementing measures under ED and EL. In parallel, the EU Commission proposed to develop EU Ecolabel criteria for photovoltaic modules.

Given this, there was interest in examining the potential synergies between the different instruments. As a result, a preparatory study was launched by the EU Commission in November 2017 on solar modules, inverters and systems, to assess ED and/or EL requirements. Unlike the standard case, in which ED/EL products are assessed independently from Ecolabel or Green Public Procurement policies, for solar photovoltaic products, the preparatory work intended to occur at the same time for the four mentioned policies. This way, the European Commission would build the evidence base in one single research process, providing supporting information to ED/EL, GPP and EU Ecolabel decision-making processes, avoiding duplicities and overburdening. The study investigated also in great detail the potential for environmental improvement, including aspects relevant to the circular economy such as reuse, repair and recycling.

To assess the environmental impacts of electricity systems and evaluate the potential benefits brought by the switch to renewables, one obvious approach is the use of life cycle assessment (LCA) [7]. It is a useful decision-support tool to quantify the environmental impacts of a product, technology or system from a life cycle perspective, i.e. from the extraction of the raw materials through to their manufacture and use up to their end of life [8]. However, to be of relevant use, a LCA study should report the values, or give an interpretation of the results per component/substance, in order to support hotspot identification. This is specifically useful to develop requirements, e.g. for EU Ecolabel.

A systematic LCA review was conducted as part of the preparatory study with a focus on the information needs of the policy tools. The LCA review analysis has complemented the identification of hotspots at component and life cycle stages, and the determination of the type of information needed to translate hotspots into verifiable criteria on aspects of performance for which there is improvement potential. LCA evidence has therefore been translated into technical performance-based criteria for the PV product group. This has been detailed in Sect. 2. For ED, it has been preliminarily identified that for modules a minimum level of energy yield and reporting on performance degradation should be achieved under fixed climatic conditions. For inverters, a minimum efficiency shall be defined, together with repairable key components. For the EU Ecolabel, it has been found that the repairability of key components along the design lifetime, as well as energy return on investment, could be feasible. Project stage-related criteria that minimize both life cycle environmental impacts and costs, together with GWP-based impact category results – as required in some national PV capacity auctions – could be integrated into a GPP criteria set. The proposals for the four policy instruments are detailed in Sect. 3.

2 Methodology

The standard preparatory studies on Ecodesign/Energy Labelling are conducted by a specific methodology for energy-related products (MEErP) [9]. Given that a combined approach between the analysis on ED/EL, GPP and the EU Ecolabel was envisaged for this specific study, additional methodological considerations were needed to complement MEErP. Moreover, the draft Product Environmental Footprint Category Rules (PEFCR) for 'Production of photovoltaic modules used in photovoltaic power systems' have been a complementary source for the identification of environmental hotspots for photovoltaic modules [10].

For its practical operation, the current version of the MEErP makes use of the so-called *Ecoreport* tool, which is a streamlined (i.e. simplified and standardized) life cycle analysis (LCA), that leads to the identification of the environmental 'hotspots' of a product or system of products, and to a quantification of the purchase cost, and production cost over the whole life cycle of the product. Once this information is available, the second part of the process (the techno-economic-environmental assessment) takes place, which takes the form of a ranking of various design options according to their life cycle costs. The analysis of the life cycle costs leads to the identification of the design option that delivers to a consumer the least life cycle cost (LLCC). The LLCC is unique per product category and provides the optimum level from a regulatory perspective because it minimizes the total cost of ownership for the consumer, and it pushes all manufacturers, at the same time, to make the necessary improvements on their products with existing technologies to produce designs linked to the LLCC.

The EU Ecolabel criteria shall among other requirements under the regulation be based on the environmental performance of products, take into account the latest strategic objectives of the community in the field of the environment and be determined on a scientific basis considering the whole life cycle of products. Compared to ED/EL, it investigates more thoroughly chemistry and toxicity aspects and tries to define the best in class based on an overall environmental assessment.

The EU GPP criteria shall mainly take into consideration the net environmental balance between the environmental benefits and burdens, including health and safety aspects. They also shall be based on the most significant environmental impacts of the product, be expressed as far as reasonably possible via technical key environmental performance indicators of the product and be easily verifiable. They also usually include a life cycle cost perspective, to encourage consideration of the total cost of ownership and not just the lowest bid price.

Figure 1 shows the overlay of EU product policy instruments under development when looking at the relative sustainability of products they target. In particular, for example, EU Ecolabel offers a higher sustainability, and GPP support for innovation through voluntary initiatives.

As prescribed by the MEErP, base cases for modules for inverters and for systems were defined.[1] The selected base case for modules is a module consisting of multicrystalline silicon cell back surface field (BSF) design, later updated to a multicrystalline silicon cell PERC (passivated emitter rear cell) design to reflect advancements in market share. For inverters, three base cases have been selected, a 2500 W string one-phase inverter, a 20 kW string three-phase inverter and a central inverter. The selected base cases for systems are a combination of the proposed base cases for modules and inverters, deployed in three types of segments: residential, commercial and utility scale with the rated capacities of 3 kW, 24.4 kW and 1.875 MW. An environmental and economic assessment of the base cases identified along the preparatory study was undertaken following the MEErP.

Then a screening of existing LCA literature has been made to identify 'hotspots' for environmental impacts along the life cycle. These may relate to specific material flows/inputs, components or emissions related to a life cycle stage. A preliminary analysis has then been made of the potential for EU Ecolabel and/or GPP criteria to

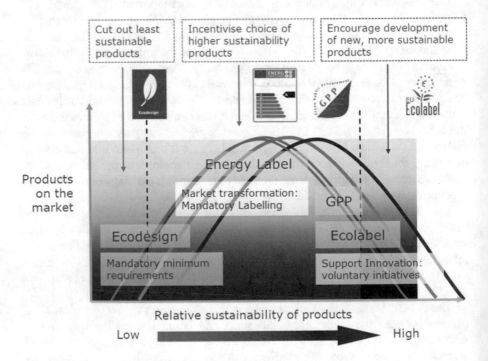

Fig. 1 Overlay of EU product policy instruments under development

---

[1] See Task 4 of the preparatory study for a detailed description of the base cases.

address these hotspots. Table 1 shows a summary of the analysis made to translate the findings from the LCA review for module inverters and systems into possible criteria.

3 Results

The focus of the preparatory study has been on the feasibility of employing four individual policy instruments, either individually or in combination. Each instrument has distinct characteristics and requirements that must be taken into consideration when deciding whether an intervention in the market is required. The proposals for each are each briefly summarized in Table 2 and presented in the sections below.

3.1 Policy Requirements Proposal: Mandatory Instruments

Policy recommendations based on the results of the analysis in the preparatory study and hotspots identification are presented below. In this context, the added value brought by each instrument and the potential synergies are considered as well as the relevance and feasibility of potentially having the product(s) covered by one or several schemes.

3.1.1 Recommendation 1: Ecodesign Minimum Mandatory Requirements for Modules and Inverters

(1) Requirements are proposed for modules on lifetime electricity yield, quality, durability, and circularity. On the yield, the preferred option is for an Ecodesign information requirement. The reason for selecting this option is that it is more representative of performance under real life conditions. The yield also takes into account PV module performance characteristics such as the spectral response under low light conditions. However, thresholds/information on the market spread for PV modules is currently missing.

(2) Another Ecodesign option could be to introduce a stringent set of quality and durability tests for module products. Testing is costly and timely; however, it is understood to already be considered as a market entry requirement by major manufacturers, and it may be difficult to separate the test sequences and/or to introduce recommended new aspects (such as encapsulant browning or inspections for cell cracking). Requirements for inverters on efficiency quality, durability and circularity are also important. The first option is based on the calculation the 'Euro Efficiency' of an inverter. This is an important derating factor for the performance of a solar PV system, so the removal of the worst performing, sub 94% efficient inverters, would contribute as a minimum

Table 1 Summary of hotspots to be translated into criteria for EU Ecolabel at R (residential), C (commercial) and U (Utility) segments

Product	PV tech/system size	Hotspots LCA	Improvement measures identified (suitability)	Scoping of the improvement potential	Possible technical requirement	Verification options	Precedents
	Si tech	Ingot/wafer production	(1) Low-energy manufacturing processes (2) Si ingot slicing, e.g. change of laser cutting, lift-off, kerfless, etc.	Reduction of: (1) Primary energy consumed (2) Losses from slicing and minimizing the Si needed for the same energy output	(1) Reduction in primary energy from ingot/wafer manufacturing (2) Reduction in GWP from silicon slicing	(1 and 2) Primary energy and GHG emissions reporting standard production specific, e.g. ISO 14064, 50001 Ener. Manag. Syst.	NSF 457 (7.1.1 required criteria)
	Si tech	Grid electricity mix	Change of site to a location with a lower grid emissions factor	Reduction of GWP up to approx. 100%	Reduction in GWP from production stage electricity use	GHG emissions reporting standard production specific, e.g. 14064	French national PV auction, GHG emissions method

Modules	All techs	Silver metallization paste	(1) Use of less silver metallization paste (2) Substitute silver by copper plating	A reduction down to 50 mg per cell is expected to be possible by 2028	Report the amount of silver per m² or per Wp of module	No standard procedure. Could be an info requirement, similar to ROHS	–
	Thin film	Metal deposition in thin films	Use of less energy-intensive step/process	Reduction of primary energy consumed by the deposition process (reduction of, e.g. toxicity impacts)	Reduction in primary energy from metal deposition processes	Primary energy reporting according to ISO5001 Energy Management System. EPBT or ERoI calculation	NSF 457 (8.1 required criteria), Blue Angel proposal
		Extraction of Cd and Te	Reduce the consumption of Cd and Te	Two CIGS manufacturers – Solar Frontier and Stеon – claim 'RoHS compliant' modules (Cd below 0.01%)	(1) Reduction of cadmium or tellurium content (2) Circular loop recovery process for semiconductor materials	(1) No standard procedure. ROHS requirement (2) Producer responsibility scheme ensuring min. recovery level, or min. recycling	NSF 457 EoL management and design for recycling and record of annual recycling and recovery rate
	Thin film	Flat glass production	Use of thinner glass, change the type, facilitate recycling or reuse	First solar series 6 has a reduced glass thickness front: 2.8 mm. Back: 2.2 mm. Environmental impact of transport reduced	(1) Glass thickness for specific grade (2) Ease of separation of lamination from glass	(1) Verification of glass specification (2) Dismantling tests to show the separation	UBA WEEE criteria: on unloading storage and handling, on preferable recycling of glass
	All techs	Lifetime and degradation	Extended lifetime and lower failure rates	Reduction of degradation rate	(1) Establish a technical lifetime according to the yield (>80% at 30y) (2) Degradation target, e.g. < 0.5%/yr	(1) Declaration made based on field data or experimental laboratory test results	–

(continued)

Table 1 (continued)

Product	PV tech/system size	Hotspots LCA	Improvement measures identified (suitability)	Scoping of the improvement potential	Possible technical requirement	Verification options	Precedents
	All techs	Energy payback time	(1) Use of less energy intensive manufacturing processes (2) Change in geographical location	(1) Rise on the energy payback time (2) Mc Si modules installed in a reference system can have 8 years or 4.31 years if they are installed in Helsinki or Sevilla, respectively	(1) To maintain an EPBT below a certain threshold for a given climate conditions (2) To include it in an energy label	No standard exists to calculate the manufacturing primary energy. Third-party verification used against EN 15804 (EPD) standard or ISO 14064 (scope 3 CO_2 emissions) for construction prod	NSF 457 (7.1.1) French national PV auction, GHG emissions calculation method
Inverters	R&C	Print board assembly	(1) Avoiding toxic elements such as Cd, Hg, Be, As, Pb and Cr (2) Pb-free soldering techniques	Hazardous substances content limitation/ improve their supply by recovery (WEEE directive)	(1) Avoiding toxic elements such as Cd, Hg, Be, As, Pb and Cr (2) Pb-free soldering techniques (3) Ease of disassembly for EoL treatments	No standards on hazardous substances in PCBs. Declaration of: (1) Substances content (targeted list (2) Lead-free content (3) Protocols for the disassembly and recycling	Ecodesign regulations for washing machines/DWs/fridges/TVs/ servers WEEE directive – PCBs≥10 cm²

Systems	R,C,U all techs					—
		Electricity demand in the supply chain of aluminium and copper production (construction stage)	Use of less or no framing and mounting structure, use of less cabling	Dual junction box design to reduce cabling and structure (e.g. 87% cable saving by Q cells), alternative frame materials or lighter structure or roof integrated PV	Amount of cabling from module/module connections Module's GWP to capture framing Integrated modules – how to credit the integration?	Feasibility uncertain: (1) Declaration of cabling material (2) GHG emissions reporting standard production specific, e.g. 14064	—
	U	BOS in thin-film technologies	Use of lighter structures or more sustainable materials	Share of the BOS in the total impact could be lower	Dual junction box design to reduce the amount of cabling and structure, or use of lighter structure, or roof integrated PV	Feasibility uncertain: (1) Declaration of cabling material (2) GHG emissions standard production specific, e.g. 14064	—
		Consumption of Cu from the electrical installation and Al from the mounting structure	Recycled content or recovery processes	Reduce consumption of Cu from the electrical installation and Al from the mounting structure	(1) Ease of dismantling and recovery (2) Recycled content	(1) Declaration of protocols of dismantling (2) Producer responsibility scheme ensuring min. recovery level, or min. recycling	—

Table 2 Proposal for product policy instruments, scope, life cycle stage and verification

Policy instrument	Stringency	Scope	Life cycle stage	Verification
Ecodesign	Mandatory	Products, packages of products	Requirements refer normally to measurable characteristics of the product (tested use stage product performance) Material efficiency requirements relating to other LC stages (e.g. repairability, durability) can be proposed, but need to be verified on the product itself Management system for design through manufacturing to be used for conformity assessment	Market surveillance is carried out at Member State level
Energy label	Mandatory	Products, packages of products	The chosen Energy Efficiency Index (EEI) shall address performance in the use stage. The EEI cannot be applied to other LC stages	Market surveillance is carried out at Member State level
EU Ecolabel	Voluntary	Can be products or services	Criteria can be set on any LC stage and include manufacturing sites/tested product performance	MS Competent Bodies verify compliance and award the label
Green Public Procurement (GPP)	Voluntary	Can be products or services	Criteria can be set on any LC stage and can include manufacturing sites, or tested product performance (link to the subject matter)	Through evidence from tenderers provided during the procurement

requirement. Introducing a standard for the minimum durability of inverters placed on the market, together with a focus on information about the repairability of the inverter, would be an important first step in extending the potential service life of inverters, particularly for those intended to be placed in outdoor environments – as failure rates can be high during the first ten years.

An additional overarching Ecodesign option would establish a standard for the collection, analysis and presentation of module and inverter life cycle data and Life Cycle Assessment (LCA) results in the EU. It could be initially on two impact categories – primary energy (GER) and Global Warming Potential (GWP).

3.1.2 Recommendation 2: Energy Label for Residential Systems

An Energy Label for solar PV systems is proposed to target the residential market segment in order to enable consumers to make an informed choice based on the performance of system designs offered by retailers and installers. It would need to be placed on the as-built rather than the monitored performance of a system.

3.2 Policy Requirements Proposal: Voluntary Instruments

3.2.1 Recommendation 3: EU Ecolabel for Residential Systems

It is proposed that a new EU Ecolabel product group is established targeted at residential systems of <10 kWp. The multi-criteria set is recommended to comprise two aspects: the package of modules and inverters and the design and installation service provided to the retail consumer. In the first approach, the criteria for modules and inverters could make use of input data from Policy Recommendations 1 (Ecodesign) and 2 (Energy Label) in order to set criteria that have an extended and stricter focus with pass/fail criteria on life cycle performance, hazardous substances and circular design. For the service approach, there would be criteria covering aspects of the service provided by system installers, e.g. the system design, or monitoring and maintenance.

3.2.2 Recommendation 4: EU Green Public Procurement Criteria for PV Systems

It is lastly proposed that a new GPP product group is established targeted at the procurement of well-designed, high-performance, long-term PV systems, and with a broader focus also on the public authority acting as a catalyst to increase local residential installations by aggregating household demand for systems and to create demand for green (solar) electricity via arrangements such as Power Purchase Agreements.

3.2.3 Combined Policy Option Recommendations

- Combined policy option 1: Mandatory instruments plus Green Public Procurement (GPP). Introduction of the two mandatory instruments would ensure a consistent focus in the market on long-term performance and circularity, acting at both component and system level. The introduction of the GPP criteria would then be to use public sector influence, in particular at regional and local level, to exploit a range of synergies with the mandatory instruments and provide guidance and criteria in three key areas:
  - The direct procurement of new solar PV systems, with reference to component performance and life cycle requirements proposed to be established under Ecodesign
  - The establishment of procurement frameworks for residential 'reverse auctions' that would facilitate an increase in residential installations, with reference to component requirements established under Ecodesign and the Energy Label

– The auction of usage rights for public assets (land and roofs) as the basis for green (solar) electricity generation, with bilateral Power Purchase Agreements as a related option

• Combined policy option 2: Voluntary instruments plus Ecodesign. While the establishment of mandatory Ecodesign requirements would establish the units of measurement and methods required for energy yield, derating factors or performance degradation, the two voluntary instruments would provide a broader means of stimulating green innovation in a coherent framework of criteria that address life cycle hotspots, focusing attention on module and inverter designs (EU Ecolabel) and on the system service 'offer' of installers (both voluntary policies).

4   Conclusions

Recommendations for policy criteria have been derived from the main MEErP study, LCA evidence and policy-specific methodologies, forming part of a preparatory study on the feasibility to apply Ecodesign, Energy Label, EU Ecolabel and GPP to photovoltaic products. The study has been made with stakeholder input. Several challenges relating to competing policy objectives and trade-offs have had to be solved by, for example, acting partially on life cycle stages. The different performance-based policy criteria have been carefully selected by prioritizing where to act, e.g. use of proxies to ensure no burden shifting. To further support the use of LCA in policy making for energy-generating products, solutions are needed to prioritize which impact categories to focus on and to reconcile the benefits and burdens of the electricity generated and other 'embodied' impacts. One solution could be to use weighting and normalization factors as recommended under the PEF method. However, to date, no methodology exists to consistently assess the environmental burden or benefits caused by electricity generation within the context of the entire global, regional or national footprint caused by humans. If this was to become available, this information can be expected to be provided a significant support to policy making.

References

1. Directive 2009/125/EC of 21 October 2009 establishing a framework for the setting of Ecodesign requirements for energy-related products
2. Regulation (EU) 2017/1369 of the European parliament and of the council of 4 July 2017 setting a framework for energy labelling and repealing Directive 2010/30/EU
3. Regulation EC 66/2010 of the European Parliament and of the Council on the EU Ecolabel
4. Public procurement for a better environment. COM(2008)400.
5. Ecodesign Working plan 2016-19. COM(2016) 773.

6. Preparatory study to assess the feasibility of applying Ecodesign, Energy Label, Ecolabel and Green Public Procurement instruments to solar photovoltaic modules, inverters and systems, at: https://susproc.jrc.ec.europa.eu/solar_photovoltaics/documents.html
7. S. Hellweg and L. Milà i Canals, Science, 2014, 344, 1109–1113
8. Hauschild, M. Z. (2005). *Environmental Science & Technology, 39*, 81A–88A.
9. See http://www.meerp.eu/documents.htm
10. Available at: https://webgate.ec.europa.eu/fpfis/wikis/display/EUENVFP/PEFCR+Pilot%3A +Photovoltaic+electricity+generation

Open Access This chapter is licensed under the terms of the Creative Commons Attribution 4.0 International License (http://creativecommons.org/licenses/by/4.0/), which permits use, sharing, adaptation, distribution and reproduction in any medium or format, as long as you give appropriate credit to the original author(s) and the source, provide a link to the Creative Commons license and indicate if changes were made.

The images or other third party material in this chapter are included in the chapter's Creative Commons license, unless indicated otherwise in a credit line to the material. If material is not included in the chapter's Creative Commons license and your intended use is not permitted by statutory regulation or exceeds the permitted use, you will need to obtain permission directly from the copyright holder.

Part V
Sustainable Methodological Solutions

Enhancing Life Cycle Management Through the Symbiotic Use of Data Envelopment Analysis: Novel Advances in LCA + DEA

Cristina Álvarez-Rodríguez, Mario Martín-Gamboa, and Diego Iribarren

Abstract The combined use of Life Cycle Assessment and Data Envelopment Analysis (LCA + DEA) arises as a growing field of research when evaluating multiple similar entities under the umbrella of eco-efficiency and sustainability. This chapter revisits a set of four recent LCA + DEA articles within the tertiary sector to explore the novel advances offered regarding the application of the well-established five-step LCA + DEA method for enhanced sustainability benchmarking. These advances – which relate to the DEA stage of the framework – include the calculation of gradual benchmarks for continuous improvement, the period-oriented benchmarking of unidivisional or multidivisional entities, and the implementation of decision-makers' preferences in the assessment. Overall, these advances further stress the suitability of using DEA to enhance the capabilities of LCA for the sustainability-oriented management of multiple similar entities.

1 Introduction

It is generally acknowledged that life cycle approaches could benefit from the combined use of other non-life cycle approaches in order to enrich decision-making processes [1]. In particular, a growing interest is found in scientific literature regarding the synergetic application of Life Cycle Assessment (LCA) and Data Envelopment Analysis (DEA) when evaluating multiple similar entities (usually

C. Álvarez-Rodríguez
Rey Juan Carlos University, Chemical and Environmental Engineering Group,
Móstoles, Spain

M. Martín-Gamboa
Department of Environment and Planning, University of Aveiro, Centre for Environmental
and Marine Studies (CESAM), Aveiro, Portugal

D. Iribarren (✉)
IMDEA Energy, Systems Analysis Unit, Móstoles, Spain
e-mail: diego.iribarren@imdea.org

© The Author(s) 2022
Z. S. Klos et al. (eds.), *Towards a Sustainable Future - Life Cycle Management*,
https://doi.org/10.1007/978-3-030-77127-0_23

257

called decision-making units, DMUs). In this regard, the symbiotic use of DEA – a linear programming methodology to calculate the relative efficiency of multiple resembling entities [2] – leads to enhance multi-criteria decision analysis by strengthening the capabilities of LCA for the eco-efficiency and sustainability management of entities.

The available reviews in the field of LCA + DEA show an increasing global interest in this area, with a growing number of case studies mainly in the primary [3] and energy [4] sectors. On the other hand, a lack of LCA + DEA studies within the tertiary sector was identified as a knowledge gap, but recently filled by a set of works addressing the sustainability-oriented management and benchmarking of retail stores as single or network (supply chain) structures [5–8]. The goal of this chapter is to explore the novel advances linked to the DEA stage of the LCA + DEA framework for enhanced sustainability benchmarking of entities by revisiting this recent set of case studies within the tertiary sector.

2   Methodology

This chapter focuses on the potentials behind the implementation – in references [5–8] – of specific DEA models that had never been used before within the well-established five-step LCA + DEA framework. As shown in Fig. 1, this LCA + DEA

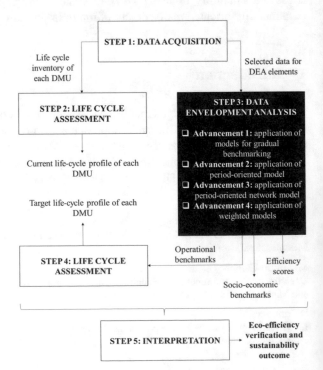

Fig. 1 Five-step LCA + DEA methodological framework and novel advancements at the DEA stage

framework involves five common stages [9]: (i) data collection for each entity under assessment (i.e., DMU) to build life cycle inventories and DEA matrices; (ii) life cycle assessment of each of the DMUs to evaluate their current life cycle profile; (iii) data envelopment analysis to compute relative efficiency scores ϕ – allowing the discrimination between efficient ($\phi = 1$) and inefficient ($\phi < 1$) DMUs – and operational and socioeconomic benchmarks (i.e., target values that would turn inefficient DMUs into efficient); (iv) life cycle assessment using life cycle inventories modified according to the operational benchmarks from the previous step, thus resulting in target life cycle profiles (or environmental benchmarks); and (v) interpretation under the umbrella of eco-efficiency and sustainability.

As mentioned above, and also highlighted in Fig. 1, the advancements reviewed in this chapter refer mainly to the DEA stage. In other words, each advancement is primarily associated with the use of specific DEA models in each original study: (i) use of DEA models for gradual benchmarking in [5], (ii) use of a period-oriented model in [6], (iii) use of a period-oriented network model in [7], and (iv) use of weighted models in [8].

Given the specific relevance of the DEA stage of the original studies, Fig. 2 shows the commonalities and singularities of these studies at this stage. Key commonalities include the inclusion of at least the store operation division for at least one annual term (year 2017) and with a common set of DEA elements. Moreover,

Fig. 2 Commonalities and singularities at the DEA stage of the revisited studies

COMMONALITIES
❖ Case study within the tertiary sector: 30 grocery stores involved
❖ Year 2017 included
❖ Electricity, receipt paper, wax paper, plastic bag, waste, and working hours as the operational and socio-economic elements of retail stores
❖ Turnover as the output of retail stores
❖ Use of input-oriented slacks-based measure of efficiency models with variables returns to scale (SBM-I-VRS)

SINGULARITIES REF. [5]	SINGULARITIES REF. [6]
❖ Use of the SBM-I-VRS and SBM-Max-I-VRS models to set a range of sustainability benchmarks for each store	❖ Use of the dynamic SBM-I-VRS model ❖ Three time terms (years 2015, 2016, and 2017) ❖ Use of economic stock as a discretionary (free) carry-over
SINGULARITIES REF. [7]	
❖ Use of the dynamic network SBM-I-VRS model ❖ Three divisions (central distribution, store operation, and home delivery) ❖ Three time terms (2015, 2016, 2017) ❖ Additional input elements (diesel in division 1, electricity in division 3, and working hours in both divisions) ❖ Additional output (home delivery service income) ❖ Use of allocated fleet and economic stock as carry-overs ❖ Use of transported merchandise as the link between divisions	**SINGULARITIES REF. [8]** ❖ Weights on DEA inputs in the case study of ref. [5]. ❖ Weights on time terms in the case study of ref. [6]. ❖ Weights on divisions in the case study of ref. [7]. ❖ Weights from the standpoint of company managers, environmental policy-makers, and local community

all these studies use input-oriented slacks-based measure of efficiency models with variables returns to scale (SBM-I-VRS), pursuing a reduction in the DEA inputs' levels while at least maintaining the same desirable output level. However, each study uses a specific SBM-I-VRS variant [10–13], which arises as a key singularity of each study: (i) use of both the conventional static SBM-I-VRS model and the alternative static SBM-Max-I-VRS model in [5] for the computation of gradual operational and socioeconomic benchmarks of retail stores, (ii) use of the dynamic SBM-I-VRS model in [6] for period-oriented sustainability benchmarking of retail stores, (iii) use of the dynamic network SBM-I-VRS model in [7] for period-oriented sustainability benchmarking of retail supply chains, and (iv) use of weighted SBM-I-VRS models/matrices to implement weights on DEA elements, time terms, or divisions according to decision-makers' preferences from the standpoint of company managers, environmental policy-makers, or local community.

It should be noted that, even though the focus is placed on the DEA stage of the five-step LCA + DEA framework, the different operational benchmarks from the DEA step directly affect the calculation of the environmental benchmarks in the fourth step and therefore the sustainability outcome of each study. Further details on the novel potentials behind each study are provided in Sect. 3.

3 Results and Discussion

Table 1 summarizes the main potentials associated with each of the studies reviewed. As a key potential linked to the use of both the conventional SBM-I-VRS model [10] and the alternative SBM-Max-I-VRS model [11], gradual sustainability benchmarking refers to the calculation – at the DEA stage – of a range of operational and socioeconomic target values (i.e., benchmarks) for each inefficient DMU. Furthermore, these gradual operational benchmarks are subsequently translated into environmental benchmarks through LCA (fourth step of the methodological framework). The computation of gradual sustainability benchmarks avoids pursuing too ambitious target values from the beginning, rationing the pursuit of efficiency and thereby promoting continuous improvement practices.

As another key potential – in this case linked to the use of the dynamic SBM-I-VRS model [12] – period-oriented sustainability benchmarking means the

Table 1 Main potentials of the novel advancements identified in LCA + DEA

Source	Novel LCA + DEA potential
[5]	Gradual sustainability benchmarking for continuous improvement
[6]	Period-oriented sustainability benchmarking
[7]	Network sustainability benchmarking for complex structures such as supply chains
[8]	Effective implementation of decision-makers' preferences (weights)

calculation, for each inefficient DMU, of operational, socioeconomic, and environmental benchmarks not only for a time term but to a number of time terms with a continuity condition between consecutive terms [14]. This allows taking into account efficiency changes over time, adapting sustainability management accordingly. Furthermore, when the DMUs are multidivisional (e.g., retail supply chains) and therefore a (dynamic) network model is used [13], this is specifically called (period-oriented) network sustainability benchmarking, as a distinction from the (period-oriented) sustainability benchmarking of unidivisional DMUs such as retail stores. The consideration of a network structure allows analysts to address the management of potentially complex entities involving interconnected processes, herein understood as divisions.

The last potential addressed in this chapter refers to the feasibility (and advisability) of implementing decision-makers' preferences (i.e., weights) in LCA + DEA studies. In this sense, the direct involvement of decision-makers such as company managers and policy-makers in an LCA + DEA study arises as a valuable asset. In fact, when decision-makers are effectively involved in the analysis, the use of weighting approaches – in addition to the default approach of equal weights – is highly recommended [8].

Finally, Table 2 summarizes the main conclusions and/or recommendations drawn from the novel LCA + DEA studies revisited in this chapter. Overall, the state of the art in LCA + DEA offers a wide range of opportunities for the sustainability-oriented management and benchmarking of multiple similar entities, fully aligning this symbiotic methodological framework with the most relevant international initiatives such as the United Nations' Sustainable Development Goals (e.g., SDG 12 on sustainable consumption and production patterns) [15] and the European Green Deal (e.g., reducing the risk of greenwashing) [16]. Moreover, further room for new potentials is still expected, which is closely linked to the wide range of life cycle approaches and DEA models available now and in the future [1].

Table 2 Main conclusions and recommendations from novel LCA + DEA studies

Source	Main conclusions/recommendations
[5]	High applicability of the LCA + DEA methodology to the service sector Feasibility of using the SBM-Max model within the LCA + DEA framework as a useful tool for gradual multidimensional benchmarking of resembling entities for continuous improvement
[6]	Suitability of the LCA + DEA methodology for period-oriented sustainability management and benchmarking of similar entities
[7]	General recommendation of enriching LCA + DEA studies by moving from unidivisional DMUs to multidivisional ones
[8]	General recommendation of enriching conventional LCA + DEA studies (which use equal weights by default) by implementing preferences from the decision-makers involved in the analysis

4   Conclusions

The novel advances explored in this chapter contribute to further strengthening the symbiosis between LCA and DEA, providing valuable general recommendations in this growing field of research. Hence, these advances are expected to boost the applicability of LCA + DEA for enhanced life cycle management, e.g., at the company level. Finally, although these advances lead to increase the interest in LCA + DEA, a high number of potentials – at the level of both methodological choices and case studies addressing new DMU categories – still remain to be unveiled.

Acknowledgments   This research has been partly supported by the Spanish Ministry of Economy, Industry and Competitiveness (ENE2015-74607-JIN AEI/FEDER/UE). Dr. Martín Gamboa states that thanks are due to FCT/MCTES for the financial support to CESAM (UID/AMB/50017/2019) through national funds.

References

1. Berlin, J., & Iribarren, D. (2018). Potentials and limitations of combined life cycle approaches and multi-dimensional assessment. In *Designing sustainable technologies, products and policies* (pp. 313–316).
2. Cooper, W. W., Seiford, L. M., & Tone, K. (2007). *Data envelopment analysis: A comprehensive text with models, applications, references and DEA-solver software*. Springer.
3. Vázquez-Rowe, I., & Iribarren, D. (2015). Review of life-cycle approaches coupled with data envelopment analysis: launching the CFP + DEA method for energy policy making. *Scientific World Journal, 813921*.
4. Martín-Gamboa, M., Iribarren, D., García-Gusano, D., & Dufour, J. (2017). A review of life-cycle approaches coupled with data envelopment analysis within multi-criteria decision analysis for sustainability assessment of energy systems. *Journal of Cleaner Production, 150*, 164–174.
5. Álvarez-Rodríguez, C., Martín-Gamboa, M., & Iribarren, D. (2019). Combined use of data envelopment analysis and life cycle assessment for operational and environmental benchmarking in the service sector: A case study of grocery stores. *Science of the Total Environment, 667*, 799–808.
6. Álvarez-Rodríguez, C., Martín-Gamboa, M., & Iribarren, D. (2019). Sustainability-oriented management of retail stores through the combination of life cycle assessment and dynamic data envelopment analysis. *Science of the Total Environment, 683*, 49–60.
7. Álvarez-Rodríguez, C., Martín-Gamboa, M., & Iribarren, D. (2020). Sustainability-oriented efficiency of retail supply chains: A combination of life cycle assessment and dynamic network data envelopment analysis. *Science of the Total Environment, 705*, 135977.
8. Álvarez-Rodríguez, C., Martín-Gamboa, M., & Iribarren, D. (2020). Sensitiy of operational and environmental benchmarks of retail stores to decision-makers' preferences through data envelopment analysis. *Science of the Total Environment, 718*, 137330.
9. Vázquez-Rowe, I., Iribarren, D., Moreira, M. T., & Feijoo, G. (2010). Combined application of life cycle assessment and data envelopment analysis as a methodological approach for the assessment of fisheries. *International Journal of Life Cycle Assessment, 15*, 272–283.
10. Tone, K. (2001). A slacks-based measure of efficiency in data envelopment analysis. *European Journal of Operational Research, 130*, 498–509.

11. Tone, K. (2016). Data envelopment analysis as a Kaizen tool: SBM variations revisited. *Bulletin of Mathematical Sciences and Applications, 16*, 49–61.
12. Tone, K., & Tsutsui, M. (2010). Dynamic DEA: A slacks-based measure approach. *Omega, 38*, 145–156.
13. Tone, K., & Tsutsui, M. (2014). Dynamic DEA with network structure: A slacks-based measure approach. *Omega, 42*, 124–131.
14. Martín-Gamboa, M., & Iribarren, D. (2016). Dynamic ecocentric assessment combining emergy and data envelopment analysis: Application to wind farms. *Resources, 5*, 8.
15. https://sustainabledevelopment.un.org/sdgs. Accessed 20.02.2020.
16. https://ec.europa.eu/info/sites/info/files/european-green-deal-communication_en.pdf. Accessed 20.02.2020.

Open Access This chapter is licensed under the terms of the Creative Commons Attribution 4.0 International License (http://creativecommons.org/licenses/by/4.0/), which permits use, sharing, adaptation, distribution and reproduction in any medium or format, as long as you give appropriate credit to the original author(s) and the source, provide a link to the Creative Commons license and indicate if changes were made.

The images or other third party material in this chapter are included in the chapter's Creative Commons license, unless indicated otherwise in a credit line to the material. If material is not included in the chapter's Creative Commons license and your intended use is not permitted by statutory regulation or exceeds the permitted use, you will need to obtain permission directly from the copyright holder.

Carbon Footprint as a First Step Towards LCA Usage

Wladmir H. Motta

Abstract In order to reduce the current intensive and inefficient use of resources and especially the negative impacts on the environment, some initiatives have emerged in different areas. Life Cycle Assessment (LCA) has been one of the most accepted and used methodology. Despite this fact, there are countries where LCA is not yet fully implemented. On the other hand, there is another approach, the carbon footprint (CF), that can follow the same life cycle approach patterns considering the phases and steps of a LCA. In this sense, this study proposes CF use as an introductory methodology of the life cycle thinking in companies at countries where LCA is still not effectively in use. The proposal is conducted through a bibliographic study and a field research. The findings point to acceptance of the proposal, considering that with the use of CF, the companies will come to know and use the principles of life cycle thinking, thus facilitating the understanding and the implementation of LCA.

1 Introduction

The continued use of natural resources at rates above the planet's regenerative capacity, mainly due to production and consumption, has brought our ecosystem to a reality of unprecedented fragility. In this sense, human activities have caused negative impacts on the environment at all scales.

Among the various evidences, those related to the various parameters of the Earth system where changes are leading the Earth system away from the relative equilibrium it had known since the beginning of the Holocene can be highlighted, and there is now discussion about the use of the term Anthropocene to specify the changes in the Earth system caused by the human species in a planetary scale, taking into account the impact of the accelerated accumulation of greenhouse gases on climate and biodiversity and also the irreversible damage caused by the overconsumption of natural resources, among others [1].

A fact that reinforces this concern is the understanding that there are nine environmental boundaries, which, once overcome, can generate severe and nonlinear

W. H. Motta (✉)
CEFET-RJ, Rio de Janeiro, Brazil

© The Author(s) 2022
Z. S. Klos et al. (eds.), *Towards a Sustainable Future - Life Cycle Management*,
https://doi.org/10.1007/978-3-030-77127-0_24

changes on the continental and planetary scale. Some of these boundaries have already been extrapolated, such as climate change, loss of biosphere integrity, changes in the terrestrial system and changes in the biogeochemical cycles of phosphorus and nitrogen [2, 3].

Another alarming data was released recently by the Intergovernmental Panel on Climate Change (IPCC), where contrary to what was expected in the face of the Paris agreement, which promised a radical transformation in technologies, investments and consumption modes, new and severely worrying data from this latest study published in 2018 (Global Warming of 1.5) exposes that the huge effort to stop global warming must be carried out immediately, precisely from 2020, or the consequences will be catastrophic [4].

Faced with the challenges posed by the ecological urgency presented, some movements emerged, such as the Paris Agreement and the Sustainable Development Goals (SDG), agreements that will require innovative approaches and contributions from all, in this sense, specifically as organizations; they started to use environmental management practices, being one of the most usual ways to initiate these practices through certifications, among which is ISO 14000.

ISO 14000 deals with the need to adapt to any change in environmental conditions, and it embodies a life-cycle approach to address these environmental aspects; among the norms of this set of norms are those referring to the carbon footprint and the life cycle assessment. Among these two proposals, life cycle assessment (LCA) is considered a valuable tool in environmental sustainability for the industry, when reviewing the complex interaction between environmental aspects and the product life cycle, being today recognized as one of the main and most comprehensive environmental tools/methodologies.

However, the dissemination of the use of this methodology is not uniform in the world, and many countries still do not use it fully; on the other hand, there is the other methodology, the carbon footprint (CF), which presents characteristics similar to LCA and brings less complexity in its implementation and may be a way to start implementing life cycle thinking in organizations.

To summarize, this chapter points out the following: (i) carbon footprint and LCA assess environmental impacts during the life cycle of products/services. The first is based on a mono-category assessment (only those related to climate change) and the second with a broader approach (multi-category based), both pointing impacts not only during the production process but also during extraction of inputs, use and end of use of products. (ii) Carbon footprint can be a first step on implementing LCA in companies. The findings point to a possibility of considering the use of the carbon footprint as a first stage in the implementation of the LCA, considering that with the use of CF, the companies will come to know and use the principles of life cycle thinking, thus facilitating the understanding and the implementation of LCA.

2 Mono- and Multi-category Assessment

In the recent past, proposals related to the reduction of environmental impacts were focused on the internal perimeter of companies, but according to current initiatives, based on the life cycle, this focus started to be supported in all phases, from the extraction of raw materials to transport, production and consumption, including final disposal and reuse. This seeks to reduce and even eliminate environmental impacts throughout the life cycle.

The life cycle assessment methodology seeks to improve the performance and environmental sustainability of production systems by providing detailed information with a view based on life cycle thinking. LCA has become a key element of environmental policies or voluntary actions in countries of the European Union, the United States, Japan, Korea, Canada, Australia and among emerging countries, such as India and, recently, China [5]. But this reality is not replicated in other countries, leaving aside, mainly developing countries.

For the United Nations Environment Program (UNEP) [6], the concept of life cycle thinking considers obtaining reliable information on environmental, social and economic impacts and makes this information available to decision-makers. It thus offers a way to incorporate sustainability into decision-making processes. It can be considered that among the various barriers related to LCA studies, the complexity of its preparation, thus consuming a lot of resources and time, is one of its main obstacles.

LCA is a multi-category methodology, as it is based on different categories of environmental impact to carry out its assessment and thus verify the necessary trade-offs, according to the options made. But in addition to this more robust and complex methodology, there are others that can be called mono-categories. This is the case for the carbon footprint that is based on only one impact category, that of greenhouse gas (GHG) emissions, related to global warming. This methodology provides reliable information on this impact, as in the case of LCA, on the life cycle.

The carbon footprint is a relatively new field of study. Its predecessor was the ecological footprint that is a measure of resource use and determines how much land area is needed to maintain a given population indefinitely [7]. The carbon footprint, however, appeared in the literature later, as described by [8], when it became more widely accepted that greenhouse gas emissions need to be reduced to avoid overheating the planet. Carbon footprint (CF) has quickly become a widely accepted term to further stimulate consumers' growing concern about issues related to climate change, being the instrument used to describe GHG emissions [9].

2.1 Standards Related to LCA and CF

Among the standards, ISO 14000 standard was initially developed with proposals for standards that organizations would follow to minimize the harmful effects on the environment generated by their activities [10]. Like ISO 9000, ISO 14000 also provides practical implementation of criteria, which includes plans aimed at making decisions that favour the prevention or mitigation of environmental impacts. The standard of management of the system in families of norms establishes requirements to direct the organization of processes that influence quality (ISO 9000) or processes that influence the impact of the organization's activities on the environment (ISO 14000).

ISO 14000 represents a voluntary international environmental standard that focuses on the structure, implementation and maintenance of an environmental management system in order to motivate organizations to systematically address the environmental impacts of their activities and establish a common approach to the challenges imposed by the ecological urgency experienced [10].

ISO 14001 standard establishes the organization's environmental management system and thus [10]:

- Promotes the assessment of the environmental consequences of the organization's activities
- Seeks to meet society's demand
- Determines policies and objectives based on the environmental indicators defined by the organization (they can portray needs from the reduction of pollutant emissions to the rational use of natural resources)
- Results in cost reduction, service provision and prevention
- Is applied to activities that may affect or affect the environment
- Is applicable to the organization as a whole

The ISO 14040 series of standards describes the principles and structure of a life cycle assessment [11]; in this sense, ISO 14044 specifies requirements and provides guidelines for LCA. As pointed out by [12], these standards include the definition of the purpose and scope of the LCA, the life cycle inventory analysis (LCI) phase, the life cycle impact assessment phase, the life cycle interpretation phase, communication and critical review of the LCA, the limitations of the LCA, the relationship between the phases of the LCA and considerations for using value choices and optional elements.

In reference to the carbon footprint, the first standard that defined it was the Green House Gas Protocol (GHG Protocol) [13], an initiative that originated in 1998, which brings together members of academia, governments and NGOs, under the coordination of the World Business Council for Sustainable Development (WBCSD) and the World Resources Institute (WRI).

The GHG Protocol formed the basis for most other carbon footprint standards. There are currently three highlighted standards for calculating the carbon footprint: ISO 14067:2018; GHG Protocol Product Life Cycle Accounting and Reporting

Standard (World Resources Institute and the World Business Council for Sustainable Development); and PAS 2050:2011 specification for the assessment of the life cycle greenhouse gas emissions of goods and services, developed by the British Standards Institution (BSI).

As for the carbon footprint normalized by ISO, in addition to ISO 14067, there are two other standards that were initially presented in 2006, namely, ISO 14064 and ISO 14065. ISO 14064, management of GHG emissions and removals, establishes standards for the quantification, monitoring and verification/validation of GHG emissions, while ISO 14065 addresses the requirements for GHG project validation and verification organizations [14]. ISO 14067: 2018 was based on the current ISO standards related to life cycle assessments (ISO 14040, ISO 14041, ISO 14042, ISO 14043 and ISO 14044) for the details for quantification, on standards related to environmental labels and statements (ISO 14020, ISO 14024 and ISO 14025) for the formatting for communication, specifies principles, and on requirements and guidelines for the quantification and communication of a product's carbon footprint [18]. This being the closest standard to ISO standards related to LCA.

For a world that continues to face this ecological urgency, organizations must continue/start to recognize the need to manage their environmental challenges and contribute to finding solutions to this common problem. Thus, the use of organizations of methodologies such as CF and LCA is very important in the face of this enormous challenge.

3 Methods and Data

This theoretical chapter aims at investigating the relationship between LCA and CF. Based on input from the literature on LCA and CF, the available evidence for this relationship was analyzed in the context of using CF as a predecessor to LCA implementation as a first step towards effective application introducing life cycle thinking. To structure the debate, a conceptual approach was carried out, and a field research on international researchers' and practitioners' perceptions on the potentially of the proposal to have CF as a first step to LCA usage will be presented.

3.1 Illustrative Case: Testimony of Experts

To add to the debate on the potentially positive use of the CF as a predecessor of the LCA, an illustrative case on international researchers' and practitioners' perceptions on this proposal will be presented.

3.1.1 Data Collection and Sample

Data collection is aimed at identifying the following aspects (among others): the state of the art of the LCA and the relationship between CF and LCA. For this purpose, a survey was designed which was disclosed and submitted through the LC Net, November–December 2015 edition, the newsletter of the Life Cycle Initiative. SurveyMonkey was used – an online survey development cloud-based software, which provides customizable surveys – for the data collection via web. The survey consisted of 15 questions organized in 9 categories according to the aspects being investigated. For the purposes of this chapter, though, it will discuss only the data related to the relationship between CF and LCA, one of the categories presented at this survey.

The questions covering this topic were structured as open questions and are composed of two questions that sought to understand at what stage is the use of the carbon footprint and validate the proposal that this can be a tool to promote the use and dissemination of LCA.

The Life Cycle Initiative was chosen to be the channel to access international researchers and practitioners with experience on LCA as it is regarded as a worldwide influential organization on the issues concerning LCA practices and its dissemination. At the time of data collection, November–December 2015, 106 Life Cycle Initiative members participated on the survey. The number of international respondents and the scope of their place of work/origin in 31 countries expressed a higher frequency of European countries with 67.0%, followed by North America with 16.0%. Regarding the time of experience, the verified distribution demonstrated a maturity of the researchers/professionals who participated, since 66.0% of the respondents had more than 6 years of experience with LCA.

3.2 Survey Responses

There were two questions on the questionnaire considering this topic. The first asked about the use and the way of using the carbon footprint in countries, seeking to understand if the methodology was already effectively used and if it would be a feasible option and already used as a first stage before the LCA. 99 responses were received: 65 (65.7%) were positive regarding the widely use of the carbon footprint, 4 (4.0%) did not know how to position themselves and 30 (30.3%) were negative concerning its use (e.g. Fig. 1). Of the 31 countries whose specialists participated in the survey, only 2 did not use the carbon footprint effectively.

The second research question was related to the proposal to use the carbon footprint as a facilitator and first step towards the dissemination of LCA practice. 97 responses were received: 71 (73.2%) were positive; 15 (15.5%) had restrictions on the LCA being more complete and requiring more details in its execution, in addition to presenting restrictions on the use of the carbon footprint as a decision tool; 2

Fig. 1 Usage of carbon footprint in countries

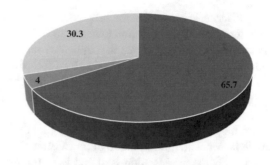

30.3

4

65.7

■ widely use of CF   ■ there is no widely use of CF   ■ did not position themselves

Fig. 2 Use of the carbon footprint as a first step towards the dissemination of LCA practice

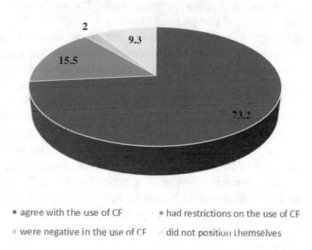

2

9.3

15.5

73.2

■ agree with the use of CF   ■ had restrictions on the use of CF
■ were negative in the use of CF   did not position themselves

(2.0%) were negative regarding its use as a first step in implementing an LCA; and 9 (9.3%) were unable to position themselves (e.g. Fig. 2).

The comments received on this proposal to use the carbon footprint as a first step towards the effective implementation of LCA were divided into three groups. One group presented positive comments on the proposals, consisting of 15 placements; another group with 9 placements presented what could be improved after the execution of the carbon footprint. The third and last group, with 14 comments, criticized the use of the carbon footprint as a precursor to LCA.

A compilation of the positive comments regarding the use of CF as a precursor to the LCA is that when conducting a carbon footprint assessment, companies come to better understand direct and indirect emissions; they come to better understand what is the approach of the life cycle and the fundamental stages of an LCA study and recognize the needs of people and resources, in addition to becoming aware of the interpretation of the results when making decisions. In this way, carbon footprint requires the execution of the most difficult parts of an LCA study, and to complement this initial study, it would be necessary to basically only collect additional data (the multi-criteria aspect) on the processes already verified in the calculation of the carbon footprint. This evolution towards an LCA study would be relatively simple.

In relation to the comments that indicated an acceptance but with a clear understanding of the differences and the needs of future actions, we have as a compilation that for companies, it is easier to start with the carbon footprint to understand the concept of LCA. The company may be frightened when faced with many categories of impact that at first may not be relevant to its products. The use of a single criterion can help for simplicity, but it involves a lot of uncertainty and choices based on a single factor. The interpretation of an LCA study is more technical due to the different impact categories addressed and is also more complex than that of the data generated by a CF.

As for the negative comments, the compilation of these positions points to a concern that a complete LCA study is more complex than just an accounting of GHG gases, made by a CF. The use of the carbon footprint may limit the understanding and scope of environmental issues in companies, making matters that are extremely complex really simplistic. Companies that perform a CF may not fully understand the concept of the life cycle and may be satisfied with just this study without understanding that they can do more through an LCA study. In the survey, 81.82% of respondents reported that the tool is used in their countries of residence/professional practice, a scope that covers 28 of the 30 countries involved in the research. The proposal to use the carbon footprint as a precursor to the LCA was accepted by the community of researchers/international experts with an approval of 73.20% of the respondents and a perceived concern on the part of 15.46% of the respondents regarding a possible loss of perception of the advantages of using the LCA methodology.

4 Discussion

In the survey, 81.82% of respondents reported that the tool is used in their countries of residence/professional practice, a scope that covers 28 of the 30 countries involved in the research. The proposal to use the carbon footprint as a precursor to the LCA was accepted by the community of researchers/international experts with an approval of 73.20% of the respondents, with a perceived concern on the part of 15.46% of the respondents regarding a possible loss of perception of the advantages of using the LCA methodology.

This concern is due to the fact that because the carbon footprint is mono-category, it verifies the impacts related only to its category (GHG emissions/global warming) and provides unilateral decision-making aimed at reducing the environmental impacts related to this category and that may eventually promote other impacts not perceived by the tool (since they are not evaluated by the tool). This fact does not occur with the LCA methodology, since it measures the impacts related to a considerable group of different categories and is able to provide information on the trade-offs that will occur due to the decisions taken with reference to these evaluated categories.

As a result of these respondents' cautious positions and positive opinions regarding the proposal, 38 comments were analysed, and from these it can be concluded that according to what was reported in the survey, the carbon footprint, although simpler than the LCA, brings the life cycle approach, its methodology and its steps into companies and can collaborate as their first contact with this approach model; the carbon footprint provides insight into the impacts generated and their dimensions for companies; the use of the carbon footprint becomes a facilitator as the life cycle study is carried out for only one impact category.

As negative aspects pointed out, several of them are relevant and are presented here: there is a need for other knowledge besides those related to the impacts responsible for global warming to be acquired and present when carrying out the LCA study; the possible difficulty in conducting the interpretation of the LCA study when carried out by the company that initially only conducted carbon footprint studies, due to the trade-offs visualized and glimpsed in as a result of the LCA studies; the fear that the methodology used to execute the carbon footprint is based on the GHG protocol or PAS 2050, which could distance the company from understanding the life cycle approach and the use of the LCA methodology; concern was shown for small businesses that would not be able to afford the costs of an LCA study; limitations regarding the need to use software for LCA studies when, for carbon footprint studies, they are not necessary; and concerns about the possibility that after the use of the carbon footprint the use of the LCA may be disowned.

The carbon footprint, being considered an integral part of an LCA study, follows the same pattern (when based on ISO 14067) of the life cycle approach as the phases and steps to be followed in its application, thus bringing the practice of the life cycle approach to the companies that execute it. Another issue regarding the use of the carbon footprint as a first step in the implementation of the LCA is that this methodology, mainly due to the results and commitments assumed by the countries participating in COP 21, tends to have greater use and eventual collection, even legal, in these countries.

5 Conclusion

The carbon footprint, being considered an integral part of an LCA study, follows the same pattern (when based on ISO 14067) of the life cycle approach as the phases and steps to be followed in its application, thus bringing the practice of the life cycle approach to the companies that execute it. Another issue regarding the use of the carbon footprint as a first step in the implementation of the LCA is that the CF methodology, mainly due to the commitments assumed by the countries participating in the COP 21, Paris Agreement, tends to have greater interest and use in the countries signatories to the agreement.

The use of the carbon footprint also directly corroborates other objectives to be achieved by nations, referring here to the Sustainable Development Goals (SDGs). Among the 17 objectives assumed, the carbon footprint has a direct relationship, especially with the thirteenth objective – "Take urgent measures to combat climate change and its impacts", in addition to having other interfaces with others of the 17 objectives.

The concern reported in the survey by a portion of the respondents, regarding a possible replacement of the LCA by the carbon footprint, should be considered, but the purpose of this study is not to propose CF use as the main methodology, but to enable companies to have contact and experience with the life cycle approach, and from this first experience, they can evolve to the admittedly more complete methodology which is the LCA.

Thus, the present study suggests that the carbon footprint should be considered as a methodology to be used as a precursor to LCA studies in companies, a factor that tends to facilitate a comprehensive implementation of LCA in countries where this practice is not yet a reality. It is hoped that this study can motivate more in-depth research and practical applications that can reinforce the pointed interrelation and proposal.

References

1. Issberner, L.-R., & Lená, P. (2018). *Anthropocene: The vital challenges of a scientific debate*. In The UNESCO Currier, Abril/June.
2. Rockstrom, J., & Steffen, W. (2009). Planetary boundaries: Exploring the safe operating space for humanity. *Ecology and Society, 14*(2), 32.
3. European Commission EUROPE 2020: A Strategy for Smart, Sustainable and Inclusive Growth, Brussels, 3.3.2010. Communication from the Commission, COM (2010/ 2020), 2010.
4. IPCC. (2018). Global Warming of 1.5 °C., in: https://www.ipcc.ch/sr15/.
5. Guinée, J. B., Heijungs, R., & Huppes, G. (2011). Life cycle assessment: Past, present and future. *Environmental Science & Technology, 45*(1), 90–96.
6. United Nations Environment Programme – UNEP/SETAC, Greening the economy through life cycle thinking: Ten years of the UNEP/SETAC Life Cycle Initiative, Paris, 2012.
7. Wackernagel, M., & Rees, W. E. (1996). *Our ecological footprint reducing human impact on the Earth*. New Society Publishers.
8. Wiedmann, T., & Minx, J. (2008). Chapter 1: A definition of 'Carbon Footprint'. In C. C. Pertsova (Ed.), *Ecological economics research trends* (pp. 1–11). Nova Science Publishers.
9. Esty, D. C., & Winston, A. S. (2008). *O verde que vale ouro*. Elsevier.
10. ISO. ISO 14001:2015 Environmental management systems – Requirements with guidance for use, 2015.
11. ISO. ISO 14040, Environmental management – Life cycle assessment – Principles and framework. Geneva, Switzerland, 2006.
12. Palma-Rojas, S., Paiva-Castro, P., Gama-Lusta, C., & Lamb, C. R. (2012). *Sistema brasileiro de inventário de ciclo de vida (SICV Brasil) e a ISO 14.044:2009*. In Congresso Brasileiro em Gestão do Ciclo de Vida de Produtos e Serviços, 3., Maringá.
13. World Resources Institute and World Business Council For Sustainable Development, The greenhouse gas protocol, Technical Report, 2000.
14. ISO. ISO14067, Carbon footprint of products – Requirements and guideline, 2018.

Open Access This chapter is licensed under the terms of the Creative Commons Attribution 4.0 International License (http://creativecommons.org/licenses/by/4.0/), which permits use, sharing, adaptation, distribution and reproduction in any medium or format, as long as you give appropriate credit to the original author(s) and the source, provide a link to the Creative Commons license and indicate if changes were made.

The images or other third party material in this chapter are included in the chapter's Creative Commons license, unless indicated otherwise in a credit line to the material. If material is not included in the chapter's Creative Commons license and your intended use is not permitted by statutory regulation or exceeds the permitted use, you will need to obtain permission directly from the copyright holder.

Society's Perception-Based Characterization Factors for Mismanaged Polymers at End of Life

Ricardo Dias, Guilherme Zanghelini, Edivan Cherubini, Jorge Delgado, and Yuki Kabe

Abstract Society's perception of an environmental impact often turns it into the drive to measure, remediate and ultimately solve the perceived problem. In some cases, this situation is noticeable even before scientists can properly establish the cause-effect pathway, for example, plastic debris effect on the oceans. This work strives to understand how public opinion deals with this transitory gap of knowledge and how to measure society's viewpoint through marine litter. A Life Cycle Assessment was addressed comparing reusable and single-use drinking straws, from which a "society's perception based" characterization factor for mismanaged polymers at end of life was proposed. Results showed that the factor may reach up to 1 order of magnitude higher than the characterization factors of producing the polymer and may indicate that decisions with no data to support can lead to rebound effects.

1 Introduction

Marine litter consists of items that have been deliberately discarded, unintentionally lost or transported by winds and rivers, into the sea and on beaches [1]. Based on this concept, it is not difficult to understand why plastic products conform most of the waste found in oceans [2]. Plastic products are often incorrectly disposed [3, 4] at end of life (EoL), worsened by the lack of economic value as waste [5, 6]. In addition to collection and sorting difficulties, this economic condition discourages plastic waste flows to circulate in the current recycling schemes. Plastic are easily transported into nature due to general product characteristic, e.g. lightweight,

R. Dias · J. Delgado · Y. Kabe
Braskem S.A., São Paulo, Brazil

G. Zanghelini (✉) · E. Cherubini
EnCiclo Soluções Sustentáveis Ltda., Florianópolis, Brazil
e-mail: guilherme@enciclo.com.br

© The Author(s) 2022
Z. S. Klos et al. (eds.), *Towards a Sustainable Future - Life Cycle Management*,
https://doi.org/10.1007/978-3-030-77127-0_25

small-sized and the float potential. Consequently, plastic waste may reach oceans via inland waterways, wastewater outflows and transport by wind or tides [7].

Statistical researchers endorse this scenario. In the European Union, 80–85% of marine litter, measured as beach litter counts, is plastic, with single-use plastic items representing 50% and fishing-related items representing 27% of the total [8]. Estimates based on 192 coastal countries pinpoint that from 31.9 million MT of mismanaged plastic in 2010, 4.8 to 12.7 million MT entered the ocean [7]. However, despite the significant values raised by these references, which indicates a constant accumulation over the decades, the issue gained prominence only in the last years, when global society started to worry about the effects of marine litter, mainly due to its impacts over marine biodiversity.

The disposable plastic drinking straw may be indicated as the most representative flagship of this current society's concern. It has been a hot topic since 2015, after a video showing a drinking straw stuck in a sea turtle's nose [9]. Since then, this product turned into the image of marine litter problem and boosted by society's opinion about the situation, propelled a large movement to eliminate plastic straws from our daily lives [10–15].

There are two aspects of this situation that became clear since 2015: (a) the overall movement had positive influence on marine litter waste problem recognition and (at some extension) on directing efforts to solve it, and (b) with laws, policies and prohibitions, alternative solutions as reusable straws or specially designed cup lids that perform the same function, gained prominence. However, despite the common intention to deal with plastic waste on the oceans, alternative scenarios may suffer with trade-off conditions as pointed by [6, 16, 17], whereas the simple prohibition without the proper scientific validation may cause rebound effects in medium to long terms (e.g. increase climate change).

Life Cycle Assessment (LCA) is able to identify this trade-off conditions between different scenarios [18–21] being recognized as a trustworthy, scientific and understandable approach that uses several mathematical models to address sustainability aspects of human activities [22–24]. However, it currently lacks a marine impact focus and robust models to account for the environmental effects of leakage into the natural environment [5, 25] especially related to the Life Cycle Impact Assessment (LCIA) framework [25–27]. On top of that, current EoL scenarios dealing with plastic waste on LCA, as sanitary landfilling, are not well addressed by impact category mechanisms due to specific product characteristics (low degradability; impacts are predominantly physical but may be also biological/chemical thorough the years) and the difficulties reproducing such complex cause-effect chains in a mathematic model. As an effect in these cases, LCA can produce some asymmetry that can lead into a misleading decision-making with not carefully considered premises and critical analysis over modelling.

Consequently, there is a major gap between scientific research and the environmental technical analysis related to "what is happening" in marine ecosystems. While this bridge is not built, this gap is being fulfilled by society judgement over the theme.

Society has built an opinion about this theme based on important evidences, although empirical and anecdotal, in most cases, on the impacts plastic can cause in marine environment. However, there is still lack of scientific development to assure the real magnitude of the damage or to trace the cause-effect pathway. Nevertheless, public policies established worldwide based only in this perception may not comprise the whole picture and may be potentially subject to failure, for example, promoting environmental trade-offs between life cycle stages or different product alternatives. Thus, the aim of this paper is to provide insights to this discussion by calculating the impact factor that is addressed to the LCA score by a new impact category based on society's perception on marine litter.

2 Material and Methods

A comparative LCA ISO compliant (i.e. LCA conducted by a LCA consulting company and reviewed by an independent third-part reviewer institution) [19, 28] was performed to assess five different drinking straws, representatives of the main commercial one-way and reusable alternatives available in the Brazilian market in 2018. However, for the sake of brevity, only plastic (marine litter case related) and stainless steel (best LCA score within reusable alternatives) options are presented since they are also the base case study of this paper. Boundaries were stablished from cradle to grave for the functional unit (FU) of "to drink 300 ml of a generic liquid from a regular glass". Their main characteristics and simplified scenario scoping are presented in Table 1.

Table 1 Main characteristics of product systems under analysis

Characteristics	Plastic drinking straw	Stainless steel drinking straw
Illustrative image		
Length/diameter (cm)	21.00/0.50	20.00/0.61
Predominant material	Polypropylene (PP)	Stainless steel/304
Main material weight (g)	0.33	11.03
Packaging	Low-density polyethylene	N.A.
Packaging weight (g)	0.09	N.A.
Additional elements	N.A.	Wire-nylon brush, cotton bag
Kit weight (g)	0.42	26.87
Washing	No	Yes
Lifetime (reuses)	One way	500 uses
End of life	Sanitary landfill	Sanitary landfill

Information from [6, 29–31] and product acquisition

The foreground data regarding raw materials weight is from primary sources, measured through a gravimetric procedure by precision scale on real (acquired) products, including primary packaging and additional elements. For raw material acquisition and material transformation, data were gathered exclusively from secondary sources such as the ecoinvent® database version 3. The washing step of stainless steel straw represents a manual and domestic process, representing an average of ten processes measured in loco for water and washing agent consumptions and effluent generation. EoL flows (including straws, packaging and complimentary elements) represent raw materials consumption based on mass balances, whereas landfilling was based on secondary data from literature and ecoinvent® database version 3.

A hybrid LCIA method based on IPCC [32], CML-IA [33] and ReCiPe 2008 at the midpoint level [34] was adopted with addition to an LCI-based impact category related to land use. Normalization was based on CML-IA divided by world population for ozone depletion, photochemical oxidation and eutrophication; CML non-baseline divided by world population for acidification; CML 2 divided by world population for resource consumption; ReCiPe divided by world population for climate change; and ILCD for respiratory inorganics and an estimated factor for land use. Weighting factors were defined based on major Braskem stakeholder's opinion, including company representatives, society and external specialists. The impact categories, characterization methods and normalization (N. factors) and weighting factors (W. factors) are listed in Table 2.

From the single score (SS) LCA results, we proposed a new impact category, namely, marine litter. This category aims to represent the society perception regarding the presence of plastic debris on the oceans, and, therefore, does not represent the traditional bottom-up approach that defines, scientifically, the cause-effect pathway (LCA characterization models). The rationale in this paper's proposal considers that characterization factor could be derived from top-down strategy (Fig. 1), based on the premise that society perception on this matter is correct.

From this perception, overall LCA SS of plastic systems should be, at least, equally environmentally harmful than other alternatives. When this condition is not

Table 2 LCIA single score method (characterization, normalization and weighting)

Impact categories (category indicator)	Source method	N. factors	W. factors
Climate change (kg CO_2 eq.)	IPCC (2013) 100a	1.45E-04	170.73
Ozone depletion (kg CFC-11 eq.)	[35]	4.41E-09	101.63
Respiratory inorganics (kg PM2.5 eq.)	[36]	0.263	109.76
Photochemical oxidation (kg C_2H_4 eq.)	CML-IA	2.72E-11	109.76
Acidification (kg SO_2 eq.)	CML-IA (non-baseline)	2.99E-12	101.63
Resource depletion, water (m^3)	ReCiPe	1.73E-03	101.63
Land use ($m^2 \cdot a$)	ReCiPe	8.91E05	101.63
Resource consumption (kg Sb eq.)	CML-IA	6.39E-12	101.63
Eutrophication (kg PO_4 eq.)	CML-IA	6.32E-12	101.63

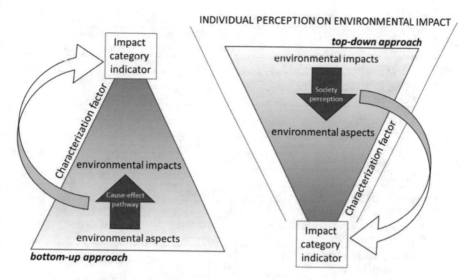

Fig. 1 Different approaches for characterization factor definition

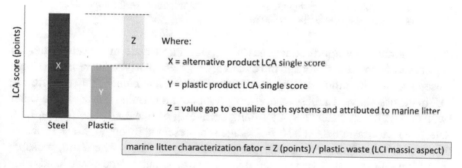

Fig. 2 Marine litter characterization factor mathematical concept (in compliance with the amount of plastic waste generated and their risk of becoming litter)

respected, the final LCA value gap is, therefore, attributed to the marine litter impact category representing society perception, as illustrated in Fig. 2.

3 Results and Discussions

3.1 Life Cycle Assessment of Drinking Straws

Each product system has a specific behaviour in terms of LCI as shown in Table 3. Plastic drinking straws (one-way product) have simple packaging, consisting of LDPE films (0.09 g) that are discarded directly during the use phase. Stainless steel

Table 3 Drinking straws Life Cycle Inventory (LCI)

	Flows[a]	Unit	Plastic drinking straw	Stainless steel drinking straw
Inputs	Polypropylene	g	0.33	–
	Stainless steel	g	–	0.022[b]
	Tin wire	g	–	9.7E-03[b]
	Nylon	g	–	5.1E-04[b]
	Cotton	g	–	0.021[b]
	LDPE	g	0.09	–
	Tap water	L	–	0.60
	Detergent	g	–	1.00
Outputs	**Drinking straw (FU)**	**p**	**1.00**	**1.00**
	Effluent (water/detergent)	L	–	0.61
	Plastic residues for treatment (sanitary landfill)	g	0.42	5.1E-04[b]
	Metal residues for treatment (sanitary landfill)	g	–	2.3E-02[b]
	Textile residues for treatment (sanitary landfill)	g	–	2.1E-02[b]

[a]Material/resource flows considering the amount of inputs/outputs to perform the FU
[b]LCI flows influenced by reuse rates 1/500 rate)

drinking straw (reusable product) has a carrying bag (made of woven cotton) to accommodate both the straw and the cleaning brush. Similarly to the reusable straw, these elements are influenced by reuse rate, having their inputs diluted to fill the FU. In the use phase, stainless steel straw presuppose a washing phase, where water (with 600 ml of tap water) and a washing agent (1 g of linear alkyl sulfonate, LAS detergent) are consumed. At last, EoL stage is represented by output flows in accordance with mass balance over the previous life cycle steps. Therefore, reusable straws have lower solid wastes than one-way straws, but on the other hand, they have a significant liquid effluent generated during the washing process (use phase).

Within the LCA scoping of this paper, single score results show a better environmental performance for plastic drinking straw with lower impacts (31.3µPt) if compared to stainless steel drinking straw (393.2µPt), as depicted by Fig. 3. Climate change, respiratory inorganics and resource depletion (water) are the main contributors to the final single score of both drinking straws with the major difference in terms of values related to the water consumption, followed by impacts due to respiratory effects.

Raw material acquisition (i.e. PP production/pellet), commonly a hotspot for one-way plastic LCA [37, 38], is the main driver for all impact categories in the case of the plastic drinking straw. Stainless steel straw has hotspots positioned mainly in additional element production (woven cotton and wired tin), detergent production and tap water consumption (during washing process). Those conditions turn the stainless steel straw into a worst environmental choice than plastic drinking straw,

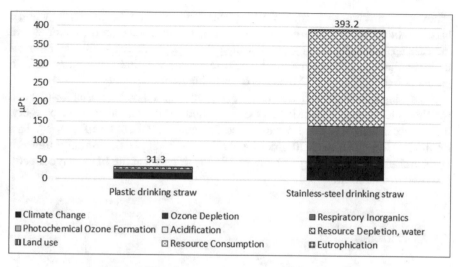

Fig. 3 Single score LCA results

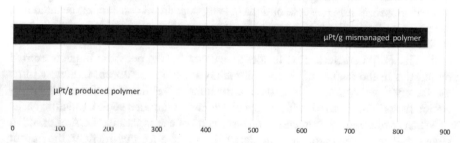

Fig. 4 Perspective of the magnitude of the new impact characterization factor considering 100% of plastic as marine litter

with its production (including mining and steel processing) and EoL being significantly diluted by reuse rate. Similar results are shown by [37, 39, 40].

3.2 Society's Perception-Based Characterization Factor

Assuming that the difference of 362 µPt between the SS results from Fig. 3 should be attributed to the plastic drinking straw final disposal flow, according to the equation in Fig. 2, we can estimate the marine litter characterization factor as 860 µPt per gram of mismanaged polymer (assuming that 100% of polymer consumption in the plastic drinking straw life cycle becomes marine litter). Comparing the impact estimated for the final disposal flow with the PP production demonstrates that this factor represents an increase of 1048% (Fig. 4). This means that the mismanaged flow represents an impact 10.5 times higher compared to the polypropylene upstream chain (i.e. equivalent to 1 order of magnitude). If we assume that 3% of the world's plastic production ends up in the oceans [7], the characterization factor would increase up to 28722 µPt per g of mismanaged plastic. In this case, the environmental impact assigned to marine litter would represent 363 times more than its own production (a difference of 2.6 orders of magnitude).

Analysing the results with a different perspective based on the normalization and weighting factors of the Braskem's LCIA method for the climate change category, the impact of plastics in the ocean would be equal to 2.2E-2 kg of CO_2 eq. This result represents an impact 20 times higher than the total plastic drinking straw life cycle emissions (1.04E-3 kg CO_2 eq. or 18 µPt) to perform the FU when correctly disposed in a landfill.

4 Conclusions

According to LCA results, polymer-based solutions tend to have better environmental performance when compared to the stainless steel reusable alternative, if correctly disposed. While reusable options heavily depend on consumers' behaviour at use phase, polymer single-use option is dependent of consumers' behaviour at end-of-life step.

The lack of characterization factors to account for the potential impacts exerted by plastics in the natural environment, mainly those in the ocean, indirectly turns the society's perception of the problem, the qualitative measure of the "characterization factor" for the marine litter impact category, without a sound scientific basis.

When attributing this perception on the results of a comparative LCA of drinking straws, following the rationale of society's perspective for marine litter, the impact of mismanaged plastics can potentially represent 10.5 and 363 times greater than its own production impacts if 100% and 3% of the plastic are considered marine litter, respectively. In both situations, the value seems to be overrated.

Other perspective, based on the climate change at midpoint LCIA level, indicates that it would be necessary 20 times more CO_2 equivalent emissions only to equalize the single score results of 0.42 g of mismanaged plastic. In both cases, LCA results due to characterization factor based on public opinion seem to be significantly higher and unbalanced with the other life cycle stages of the plastic drinking straw. Thus, society does not perceive the impacts of the polymer straw application as LCA results may indicate, mainly in order of magnitude.

Even though this work's aim is to present a case as an exercise and not to properly calculate a reproducible characterization factor, it gives insight about the current LCA gap of knowledge and how far an LCA result may be from public opinion. Doubtlessly science should not be nudged by any perception, and real characterizations factors are still to be calculated. The lack of data, high complexity of the subject, and the difficulty of proper communication between scientific community and social influencers tend to lead people to the precautionary side and to make decisions with no data to support. In this case, society becomes very prone to suffer from rebound effects.

References

1. European Commission. (2010). *Marine litter: Time to clean up our act*. https://ec.europa.eu/environment/marine/pdf/flyer_marine_litter.pdf (Accessed 23.02.2020).
2. Derraik, J. G. B. (2002). The pollution of the marine environment by plastic debris: A review. *Marine Pollution Bulletin, 44*(9).
3. Department for Environment, Food & Rural Affairs. (2018). *Our waste, our resources: A strategy for England – Evidence annex, HM Government*, DEFRA, 129 pp.
4. United Nations Environment Programme. (2018). *Single-use plastics: A roadmap for sustainability*. UNEP, 104 pp.
5. Ellen MacArthur Foundation. (2017). *the new plastics economy: Rethinking the future of plastics & catalysing action*. Ellen MacArthur Foundation, 68 pp.
6. Department for Environment, Food & Rural Affairs. (2018). *A preliminary assessment of the economic, environmental and social impacts of a potential ban on plastic straws, plastic stem cotton buds and plastics drinks stirrers*, DEFRA, 91pp.
7. Jambeck, J. R., Geyer, R., Wilcox, C., Siegler, T. R., Perryman, M., Andrady, A., Narayan, R., & Law, K. L. (2015). Plastic waste inputs from land into the ocean. *Science, 347*(6223), 768–771.
8. Directive EU, 2019/904, 2019, on the reduction of the impact of certain plastic products on the environment (Text with EEA relevance), Official Journal of the European Union, 2019, pp 19.
9. The Leatherback Trust. (2015). *Removing a plastic straw from a sea turtle's nostril – Short Version*. https://www.youtube.com/watch?v=d2J2qdOrW44.
10. Barbosa, V. (2018). Rio de Janeiro é primeira capital brasileira a proibir canudos plásticos, Revista Exame. https://exame.abril.com.br/brasil/rio-de-janeiro-e-primeira-cidade-brasileira-a-proibir-canudos-plasticos/ (20.01.2019).
11. Department for Environment, Food & Rural Affairs. (2018). *Government launches plan to ban plastic straws, cotton-buds, and stirrers*. DEFRA. https://www.gov.uk/government/news/government-launches-plan-to-ban-plastic-straws-cotton-buds-and-stirrers (21.01.2019).
12. Garrand, D. (2018). Seattle ban on plastic straws to go into effect July 1. *CBS News*. https://www.cbsnews.com/news/seattle-ban-on-plastic-straws-goes-into-effect-july-1/ (21.01.2019)
13. Gibbens, S. (2019). A brief history of how plastic straws took over the world. *National Geographic*. Vol. Environment – Planet or Plastic? https://www.nationalgeographic.com/environment/2018/07/news-plastic-drinking-straw-history-ban/ (21.01.2019).
14. Rosenbaum, S. (2018). She recorded that heartbreaking turtle video. Here's what she wants companies like Starbucks to know about plastic straws. *Time Magazine*. http://time.com/5339037/turtle-video-plastic-straw-ban/ (19.01.2019).
15. The Last Straw. (2018). http://www.laststraw.com.au/ (22.01.2019).
16. Britschgi, C. (2018). *Starbucks Bans Plastic Straws, Winds up using more plastic. A reason investigation reveals that the coffee giant's new cold drink lids use more plastic than the old straw/lid combo*. Reason – Free Mind and Free Market. https://reason.com/blog/2018/07/12/starbucks-straw-ban-will-see-the-company (20.01.2019).
17. Tarrant, H. (2018). *The Plastic Straw Dilemma is not what it seems*. Medium Corporation. https://medium.com/@creativeharm/the-plastic-straw-dilemma-4338c76269c0 (20.01.2019).
18. Baumann, H., & Tillman, A. M. (2004). *The Hitch Hiker's guide to LCA: An orientation in life cycle assessment methodology and application* (1st ed.). Studentlitteratur.
19. ISO, ISO 14040:2006, Environmental Management – Life Cycle Assessment – Principles and Framework, International Organization for Standardization.

20. Reap, J., Roman, F., Duncan, S., & Bras, B. (2008). A survey of unresolved problems in life cycle assessment. Part 1: Goal and scope and inventory analysis. *Internationational Journal of Life Cycle Assess, 13*, 290–300.
21. Von Doderer, C. C. C., & Kleynhans, T. E. (2014). Determining the most sustainable lignocellulosic bioenergy system following a case study approach. *Biomass Bioenergy, 70*, 273–286.
22. Baitz, M., Albrecht, S., Brauner, E., Broadbent, C., Castellan, G., Conrath, P., Fava, J., Finkbeiner, M., Fischer, M., Fullana, I., Palmer, P., Krinke, S., Leroy, C., Loebel, O., Mckeown, P., Mersiowsky, I., Möginger, B., Pfaadt, M., Rebitzer, G., Rother, E., Ruhland, K., Schanssema, A., & Tikana, L. (2013). LCA's theory and practice: Like ebony and ivory living in perfect harmony? *International Journal of Life Cycle Assessment, 18*, 5–13.
23. Cherubini, E., Zanghelini, G. M., Alvarenga, R. A. F., Franco, D., & Soares, S. R. (2015). Life cycle assessment of swine production in Brazil: A comparison of four manure management systems. *Journal of Cleaner Production, 87*, 68–77.
24. Finnveden, G., Hauschild, M. Z., Ekvall, T., Guinee, J., Heijungs, R., Hellweg, S., Koehler, A., Pennington, D., & Suh, S. (2009). Recent developments in Life Cycle Assessment. *Journal of Environmental Management, 91*, 1–21.
25. Woods, J. S., Veltman, K., Huijbregts, M. A. J., Verones, F., & Hertwich, E. G. (2016). Towards a meaningful assessment of marine ecological impacts in life cycle assessment (LCA). *Environment International, 89-90*, 48–61.
26. Casagrande, N. M. (2018). Inclusão dos impactos dos resíduos plásticos no ambiente marinho em avaliação do ciclo de vida, Dissertação (Mestrado em Engenharia Ambiental), Universidade Federal de Santa Catarina, Florianópolis, 113p.
27. Sonnemann, G., & Valdivia, S. (2017). Medellin declaration on Marine Litter in Life Cycle Assessment and Management. *International Journal of Life Cycle Assess, 22*, 1637–1639.
28. ISO, ISO 14044:2006, Environmental Management – Life Cycle Assessment. – Requirements and Guidelines, International Organization for Standardization.
29. Boonniteewanich, J., Pitivut, S., Tongjoy, S., Lapnonkawow, S., & Suttiruengwong, S. Evaluation of Carbon Footprint of Bioplastic Straw compared to Petroleum based Straw Products, 2014, 11th Eco-Energy and Materials Science and Engineering. *Energy Procedia,* Vol. 56, pp. 518–524.
30. Strawplast. (2018). Canudos descartáveis. http://strawplast.com.br/ (23.09.2018).
31. Beegreen, Canudo Reto, 2018., https://loja.beegreen.eco.br/canudo-reto, (23.09.2018).
32. Stocker, T. F., Qin, D., Plattner, G.-K., Tignor, M., Allen, S. K., Boschung, J., Nauels, A., Xia, Y., Bex, V., Midgley, P. M., & IPCC. (2013). *Climate change 2013: The physical science basis. Contribution of Working Group I to the Fifth Assessment Report of the Intergovernmental Panel on Climate Change* (1535 pp). Cambridge University Press.
33. Guinée JB, Gorrée M, Heijungs R, Huppes G, Kleijn R, de Koning A, van Oers L, Wegener Sleeswijk A, Suh S, Udo de Haes HA, de Bruijn JA, van Duin R, Huijbregts MAJ, *Handbook on Life Cycle Assessment: Operational guide to the ISO standards.* Series: eco-efficiency in industry and science. Kluwer Academic Publishers, Dordrecht, 2002.
34. Goedkoop, M., Heijungs, R., Huijbregts, M. A. J., De Schryver, A., Struijs, J., & van Zelm, R. (2013, May). ReCiPe 2008: A life cycle impact assessment method which comprises harmonised category indicators at the midpoint and the endpoint level. First edition Report I: Characterisation. RIVM, Bilthoven.
35. World Meteorological Organization. (2011). Scientific Assessment of Ozone Depletion: 2010, Global Ozone Research and Monitoring Project–Report No. 52, 516 pp., Geneva, Switzerland.
36. Rabl, A., & Spadaro, J. V. The Risk Poll software, version is 1.051 (dated August 2004). www.arirabl.com.
37. Dhaliwal, H., Browne, M., Flanagan, W., Laurin, L., & Hamilton, M. (2014). A life cycle assessment of packaging options for contrast media delivery: Comparing polymer bottle vs. glass bottle. Packaging systems including recycling. *International Journal of Life Cycle Assess, 19*, 1965–1973.

38. Wood, G., & Sturges, M. (2010). *Single trip or reusable packaging – Considering the right choice for the environment*. WRAP. Final report: Reusable packaging – Factors to consider.
39. Danish Environmental Protection Agency. (2018). Life Cycle Assessment of grocery carrier bags, 2018, Environmental Project no. 1985. February pp. 144.
40. Ligthart, T. N., & Ansems, A. M. M. (2007). *Single use cups or reusable (coffee) drinking systems: An environmental comparison*. TNO-2007.

Open Access This chapter is licensed under the terms of the Creative Commons Attribution 4.0 International License (http://creativecommons.org/licenses/by/4.0/), which permits use, sharing, adaptation, distribution and reproduction in any medium or format, as long as you give appropriate credit to the original author(s) and the source, provide a link to the Creative Commons license and indicate if changes were made.

The images or other third party material in this chapter are included in the chapter's Creative Commons license, unless indicated otherwise in a credit line to the material. If material is not included in the chapter's Creative Commons license and your intended use is not permitted by statutory regulation or exceeds the permitted use, you will need to obtain permission directly from the copyright holder.

Research Activities on LCA and LCM in Poland

Zenon Foltynowicz and Zbigniew Stanisław Kłos

Abstract The main goal of this paper is to present the history and actual situation in research on LCA and LCM in Poland. This task will be performed by reviewing the different activities and their results in this field, from the very beginning. The paper includes the review of the activities of LCA/LCM main research centres in Poznań (Poznań University of Technology (PUT), Poznań University of Economics and Business (PUEB)), Cracow (Polish Academy of Sciences, AGH University of Science and Technology, Cracow University of Economics), Zielona Góra (University of Zielona Góra), Bydgoszcz (UTP University of Science and Technology), Katowice-Gliwice (Silesian University of Technology), Częstochowa (Częstochowa University of Technology) and Szczecin (ZUT Western Pomeranian University of Technology). LCA/LCM researches are also performed in several smaller research groups in R&D centres. In the end of the paper, some conclusions referring to the actual situation of research on LCA/LCM, dealing with critical evaluation of the LCA/LCM centres in Poland location, issues and problems addressed, areas of the projects covered and the desired activities in the future, are presented.

1 Introduction

Environmental life cycle assessment has developed fast over the last three decades. A comprehensive review of the historical development of LCA has recently been presented by Guinée [22]. So far, a description and summary of the state of research on LCA in Poland has been made several times, for the first time in 1990 [24]. The first studies worldwide, which are currently considered as LCA, were carried out in the late 1960s and early 1970s. In the years 1970–1990, the LCA concept was

Z. Foltynowicz (✉)
Poznań University of Economics and Business, Poznań, Poland
e-mail: zenon.foltynowicz@ue.poznan.pl

Z. S. Kłos
Faculty of Civil and Transport Engineering, Poznań University of Technology, Poznań, Poland

© The Author(s) 2022
Z. S. Klos et al. (eds.), *Towards a Sustainable Future - Life Cycle Management*,
https://doi.org/10.1007/978-3-030-77127-0_26

developed with widely divergent approaches and terminologies. The 1990s brought about a remarkable increase in research activities around the world, reflecting, inter alia, the number of published LCA guides and textbooks. In 1990–2000, harmonization of methods took place, thanks to SETAC coordination and ISO standardization activities, providing a standardized framework and terminology as well as platforms for debate and harmonization of LCA methods. In addition, the first scientific journals appeared with LCA as their main subject.

2 Early Works in Poland

As a starting point, the first attempts of introduction of LCA/LCM aspects into research practice in Poland are presented. These "pre-historical" activities were connected with the implementation of life cycle frames into analysis of environmental impacts of technical objects, as it was presented in a paper focused on consideration on the usefulness of determination of environmental impacts of the machine and device existence in the life cycle [23], published in Scientific Works of PUT, series: Machines and Vehicles, in 1986 (author: Zbigniew Kłos). Among other activities, the first book on LCA-related issues by Zbigniew Kłos entitled "Environment Protection Oriented Property of Technical Objects. A Study of Valuation of Machines and Devices Influence on Environment", published by Editions of PUT in 1990 [3], and the first PhD thesis "Ecobalancing of Machines and Devices with the Example of Air Compressors", defended by Grzegorz Laskowski at Faculty of Machines and Vehicles, PUT, in 1999 (supervisor: Zbigniew Kłos), should be pointed out. Then there were in the 1990s other activities accomplished, like engagement in work activities of European LCA research groups: SETAC-Europe Workgroup on LCA and Conceptually Related Programs and SETAC-Europe Workgroup on LCA Case Studies and participation in the European Union Research Programme LCANET as well as in thhe European Union Concerted Action CHAINET (Zbigniew Kłos). More about these works were presented in publication of Kłos [25] and Adamczyk [1] working at the University of Economics in Cracow. Since then, there have been more and more publications on the subject. In addition to these two centres, which initiated the LCA research in Poland, this topic began to develop in the following scientific centres: PUEB, University of Zielona Góra, Gdynia Maritime Academy, Mineral and Energy Economy Research Institute of Polish Academy of Sciences, Central Mining Institute and Wood Technology Institute. The innovative scope of LCA research in these centres has been discussed in a number of scientific reports, among others in the review papers of Kłos [16, 25, 26, 65], Lewandowska [15, 16, 32, 39, 65] and Kulczycka [32].

This review paper characterizes individual centres, scientists working in them and the main research topics. Our goal is not to re-describe them; however short characteristics will be presented in the research part when discussing the results of the bibliometric ranking. The growing number of publications in both national and significant international journals was also pointed out in these studies. The list of publications of Polish researchers in the journals possessing impact factor already includes several dozen items. The first publication in a leading journal, IJLCA, with Kłos co-authorship appeared in 2000 [57] and subsequent completely by national authors in 2004 [40, 43]. The following years brought further publications together with the growing number of centres starting research in the field of LCA/LCM. These publications meet a growing interest as evidenced by their increasing number of citations. However, no comparative analysis of these publications has yet been carried out. The aim of this work is therefore not only the presentation of scientists from a given Polish LCA centres but also an attempt of the bibliometric analysis of Polish LCA's scientist performance. The question arises: what kind of indicators would be really useful for such analysis? Under evaluation of a paper, the three main factors, impact factor of a journal, number of citations and year of publication, seem to determine the importance of a given publication.

3 Proposed Bibliometric Method of Polish LCA's Scientist Achievement Evaluation

3.1 Methodology

The number of scientific publications and the number of journals have increased considerably in the last few years. How to find out in this thicket which are valuable and which are not worth? Some probably remember that there is Eugene Garfield who began a new era in the processes of evaluation and measurement of scientific publications with his radical invention, the Science Citation Index (SCI), which enabled the statistical analysis of large-scale scientific literature [19]. Then, several methodologies for evaluating scientific papers were proposed [54]. Early work in this field, consisting in determining the quality of the best works, as mentioned in [54], approached the qualitative dimension of the work represented by the journal's impact factor and the number of citations of the analysed works.

The quality of work should be assessed through its impact on the scientific community. With this in mind, we used the Methodi Ordinatio [54], a method in order to rank publications of Polish LCA researchers.

3.2 Methodi Ordinatio Description

Methodi Ordinatio is a multi-criteria assessment model (InOrdinatio) used to rank publications according to a set of criteria such as journal impact factor in which the paper was published, year of publication and number of citations [54]. The equation InOrdinatio (1) is applied to identify the scientific works' ranking:

$$\text{InOrdinatio} = (\text{IF}/1000) + \text{alfa} * \left[10 - (\text{ResearchYear} - \text{PublishYear}) \right] + \text{Ci} \qquad (1)$$

Where:

- IF is the journal impact factor in which the paper was published.
- alfa is the weighting factor ranging from 1 to 10, to be attributed by the researcher.
- ResearchYear is the year in which the research was developed.
- PublishYear is the year in which the paper was published.
- Ci is the number of times the paper has been cited in the literature.

The authors of the method [54] adopted the following assumptions for the equation InOrdinatio:

(a) Originally, the impact factor IF is divided by 1000 (thousand), striving to normalize its value in relation to the other criteria. We do not agree with this assumption because it depreciates this important indicator. That is why in our calculations it was assumed that we will multiply IF by 10 to give it the right rank. It is not easy to publish an article in a journal characterized by a relatively large IF. The use of the journal impact factor in academic review, promotion and tenure evaluations has been very recently discussed by McKiernan et al. [50].
(b) The equation contains a weighting factor "alfa", the value of which the researcher assigns. It can be from 1 to 10. If its value is close to 1, it means that the researcher assigns less importance to the year of publication as a criterion, .and the closer to 10, when he assigns the greater importance of this criterion.

3.3 Methodi Ordinatio Application for Analysis of Polish Authors LCA's Publications

3.3.1 Adopted Research Assumptions

The scope of the research included publications in the field of LCA by Polish specialists. Their list was established on the basis of research in the scientific community. To calculate the InOrdinatio indicator, it was decided to use publications from the period 1995–2019. In the study, year 2010 was adopted as the current turning point. For years below 2010, the value of "alpha" as 5 was arbitrarily assumed. For the present decade, the value of "alpha" was assumed to be 10, because a shorter time elapsed since the publication and this means less time to quote by the scientific community.

3.3.2 Source of Data

There are several databases from which bibliometric data can be obtained, such as WoS or Scopus. However, in this work, it was decided to use the Google Scholar database, because it indexes not only IF journals but also other scientific publications, including books that do not have IF. Thus, data on the number of citations of publication data were obtained from Google Scholar citation. Included were publications that had LCA and/or LCM in the title or keywords as well as full headings Life Cycle Analysis/Assessment as well as Life Cycle Management.

In several cases, a problem was found due to the lack of a given author's profile on the Google Scholar platform. At that case, other available sources were used.

There are several ways to determine the citation index, including SCI, JCR and SJR. The study decided to use only JCR citation indicators that were obtained from the webpages of the magazine. Annual indicators were used, although the use of so-called 5-year indicators was also taken into account; however, they are not favourable for recent publications.

3.3.3 Calculation of InOrdinatio

The modified equation InOrdinatio (2) was applied for calculation:

$$InOrdinatio = 10 * IF + alfa * \left[10 - \left(ResearchYear - PublishYear \right) \right] + Ci \qquad (2)$$

As previously justified, IF was multiplied by 10 to reflect the importance of this indicator.

4 Results and Discussion

Research is done on the base of analysis of research activities on LCA and LCM presented in details in "Bibliometric analysis of Polish LCA's scientist performance" [14]. For each leading author from a given centre, the most-read publications with at least ten citations were usually selected. In the tables presented in report [14], they were listed according to the decreasing number of citations, from the highest first. After the InO calculations, the five best publications for the centre were selected.

The results started to be presented in alphabetical order according to the name of the leading author in the given centre, with Janusz Adamczyk, as the first author considered.

In Table 1 [14] the results of InOrdinatio for authors from the University of Zielona Góra are presented. The authors began publishing in 2014, but their best publications were published in high IF journals and reach InOrdinatio above 100 with the best InOrdinatio of 147.5. The main areas of their interest are ecological

Table 1 The results of InOrdinatio for authors from the University of Zielona Góra

Order number	Publication number according to the list	IF	Number of citation	Year of publication	InOrdinatio	Ranking
1	[58]	8.050	37	2016	**147.5**	1
2	[12]	3.324	28	2015	101.24	5
3	[2]	5.901	26	2014	134.01	2
4	[9]	3.844	19	2014	107.44	3
5	[10]	5.715	15	2016	102.15	4
6	[11]	5.715	14	2016	101.15	6

Table 2 The results of InOrdinatio for authors from the Central Mining Institute

Order number	Publication number according to the list	IF	Number of citation	Year of publication	InOrdinatio	Ranking
1	[4]	3.590	119	2013	**214.9**	1
2	[5]	4.900	32	2016	111.0	2
3	[18]	5.651	27	2017	103.51	4
4	[27]	5.715	23	2016	110.15	3
5	[7]	4.601	18	2016	94.01	5
6	[60]	4.610	15	2017	81.10	
7	[6]	3.173	15	2016	56.73	

and economic aspects of reducing low emissions using the LCA technique and LCA application in the construction industry.

Wacław Adamczyk from Cracow University of Economics should be second. His publication achievements can be found in the literature list [3]; however, the lack of his Google Scholar profile makes impossible the analysis of Methodi Ordinatio. However, it should be mentioned that Adamczyk and his team is one of the precursors in promoting life cycle thinking in relation to products. Noteworthy is also the organization of several editions of the Ecology of Products conferences, which resulted in important monographs [3]. The use of the LCA method in the decision-making processes of production companies and in their product policy is currently the main scope of activity of this research group.

In Table 2 [14] the best publications of the group whose leader is Burchard-Korol are presented. The group leader while working at the Central Mining Institute has carried out extensive work on the application of life cycle assessment and eco-efficiency in mining and quarrying sectors. From 2018 (at the Silesian University of Technology, Faculty of Transport), she has been examining the importance of assessing the environmental life cycle of transport. Noteworthy is the publication [4], which already has 119 citations, which gives rather high InOrdinatio equal 214,9.

Similar research issues were carried out by Czaplicka-Kolarz (currently she works at the Silesian University of Technology, Faculty of Organization and Development). Her papers were summarized in Table 3 [14] with the best InOrdinatio equal 111.0.

Table 3 The results of InOrdinatio for Czaplicka-Kolarz from the Central Mining Institute

Order number	Publication number according to the list	IF	Number of citation	Year of publication	InOrdinatio	Ranking
1	[5]	4.900	32	2016	**111.0**	1
2	[18]	5.651	27	2017	103.51	2
3	[7]	4.601	18	2016	94.01	3
4	[6]	3.173	15	2016	56.73	4

Table 4 The results of InOrdinatio for Foltynowicz group (from Poznan University of Economics and Business)

Order number	Publication number according to the list	IF	Number of citation	Year of publication	InOrdinatio	Ranking
1	[43]	1.6	39	2004	**130.0**	1
2	[44]	1.8	19	2008	92.0	4
3	[40]	0.366	14	2004	92.66	3
4	[39]	1.6	10	2004	101.0	2

Table 5 The results of InOrdinatio for research group from Poznan University of Technology

Order number	Publication number according to the list	IF	Number of citation	Year of publication	InOrdinatio	Ranking
1	[57]	1.039	62	2000	157.39	3
2	[41]	3.148	50	2010	**171.48**	1
3	[35]	3.148	32	2010	163.24	2
4	[65]	3.988	29	2014	118.88	4
5	[45]	3.089	26	2013	116.89	5*
6	[17]	3.173	19	2016	80.73	
7	[32]	2.362	13	2011	116.82	5*
8	[59]	3.988	12	2014	101.88	
9	[34]	3.988	8	2014	97.88	

*Same rank because of very small difference

Table 4 [14] presents the achievements of Foltynowicz group from Poznan University of Economics and Business, which was the third one who started LCA in Poland. The initial research was devoted to comparative LCA analysis of industrial objects followed by the expansive works of Lewandowska. The highest rate of InOrdinatio (130.0) is attributed to exhibit paper published in 2004 [43]. Currently, the group publishes works in the field of renewable energy (see [51, 52]).

Table 5 [14] presents the achievements of the research group from PUT Poznań (Kłos, Kasprzak, Kurczewski, et al.). The authors began publishing before year 2000 [23–25]. Their best publications reach InOrdinatio above 100 with the best of 171.48. The main areas of their interest are very broad, among other life cycle thinking in small and medium enterprises and an environmental life cycle assessment of machines and devices.

Table 6 The results of InOrdinatio for Korol

Order number	Publication number according to the list	IF	Number of citation	Year of publication	InOrdinatio	Ranking
1	[27]	5.715	23	2016	**110.15**	1
2	[6]	3.173	15	2016	56.73	2

Table 7 The results of InOrdinatio for research group from Mineral and Energy Economy Research Institute of the Polish Academy of Sciences in Cracow

Order number	Publication number according to the list	IF	Number of citation	Year of publication	InOrdinatio	Ranking
1	[33]	0.79	30	2015	77.9	
2	[45]	3.089	26	2013	**116.89**	1
3	[21]	4.732	24	2017	91.32	
4	[37]	3.173	24	2016	85.73	
5	[31]	3.331	23	2016	86.31	
6	[28]	0.25	23	2004	100.5	3
7	[20]	0.153	21	2005	92.5	5
8	[30]	2.6	20	2009	96.0	4
9	[29]	1.0	19	2007	89.0	
10	[32]	2362	13	2011	116.62	2

The achievements of Korol from the Central Mining Institute who is dealing with the evaluation of environmental footprints of biopolymers are shown in Table 6 [14].

Next, the achievements of two groups, whose leaders are strong women in LCA's science, will be presented. Table 7 [14] presents the achievements of the group whose leader is Kulczycka (Mineral and Energy Economy Research Institute of the Polish Academy of Sciences in Cracow). The issues of many works are very broad, but as befits the institute in which they work, it mainly concerns LCA issues in the field of the mineral and energy industry. The tabular summary (Table 7 [14]) shows how IF affects the InOrdinatio index. Although the largest is equal to 116.89, most publications have high citation.

Table 8 [14], which presents the achievements of the group led by Lewandowska from Poznan University of Economics and Business, contains more articles than in other cases. The reason is not only the number of publications but also the fact that they are the result of extensive cooperation with other research groups. Twelve of these works have InOrdinatio above 100. The largest InOrdinatio reach values in the range 150–170. The issues of these works include both practical and methodological aspects in the field of LCA.

The next two cases present the results of groups that publish a lot, but either in Polish language or in magazines with small IF, which affects not very high InOrdinatio. Table 9 [14] presents the achievements of the Nitkiewicz team. Nitkiewicz comes from the Kraków group of Adamczyk and currently forms a group in Częstochowa (Center of Life Cycle Modeling). Research work of this group is directly related to LCA and its applicability. Group members have

Table 8 The results of InOrdinatio for research group from Poznan University of Economics and Business

Order number	Publication number according to the list	IF	Number of citation	Year of publication	InOrdinatio	Ranking
1	[55]	2.296	60	2014	132.96	4
2	[38]	2.362	53	2011	156.62	3
3	[41]	3.148	50	2010	**171.48**	1
4	[56]	3.341	40	2014	123.41	
5	[43]	1.6	39	2004	130.0	
6	[42]	2.296	35	2014	106.96	
7	[46]	2.465	35	2013	119.65	5
8	[35]	3.148	32	2010	163.24	2
9	[47]	3.324	31	2015	104.24	
10	[65]	2.296	29	2014	101.96	
11	[45]	3.089	26	2013	116.89	
12	[37]	3.173	24	2016	85.73	
13	[17]	3.173	19	2016	80.73	
14	[44]	1.8	19	2008	92.0	
15	[40]	0.366	14	2004	92.66	
16	[32]	2.362	13	2011	116.62	

Table 9 The results of InOrdinatio for research group from the Faculty of Management at Częstochowa University of Technology

Order number	Publication number according to the list	IF	Number of citation	Year of publication	InOrdinatio	Ranking
1	[61]	1.08	10	2015	**60.8**	1
2	[62]	0	7	2014	57	2
3	[53]	1.334	1	2017	43.34	3

Table 10 The results of InOrdinatio for research group from UTP University of Science and Technology in Bydgoszcz

Order number	Publication number according to the list	IF	Number of citation	Year of publication	InOrdinatio	Ranking
1	[63]	0.763	11	2017	**38.63**	1
2	[13]	1.21	9	2018	31.1	2
3	[64]	1.214	4	2018	26.14	3

published about 30 scientific works, however, mainly in Polish publishing houses, which results that only three of them have citations. This is reflected in the low InOrdinatio values.

The situation is similar in the case of Tomporowski research group from UTP Bydgoszcz. Table 10 [14] presents selected achievements of this research group, which are cited publications from indexed periodicals. Although these publications have been published in recent years, they already have citations. Other numerous publications in non-indexed periodicals affect InOrdinatio. The subject of this research is very current and focuses on various aspects of the LCA of an offshore wind farm.

In addition to the above research groups, LCA/LCM analyses are carried out in several other centres, as evidenced by the number of licenses purchased for SimaPro or GaBi computing programs, like at ZUT (West Pomeranian University of Technology, Szczecin [8]), Łódź University [48, 49] and COBRO Institute [36, 66, 67].

5 Reassuming and Conclusions

The bibliometric analysis of Polish LCA's scientists' performance has been performed. Based on the review of discipline-related journals and the information collected, InOrdinatio was determined using the Methodi Ordinatio. The year of publication and the number of citations of the publication were taken into account, as well as the IF of the magazine in which the article was published. On this basis, InOrdinatio was determined, and the best five publications from a given centre were

Table 11 Ranking of the best papers from Polish LCA research groups

Ranking number	Publication number[a]	IF	Number of citation	Year of publication	InOrdinatio	InOrdinatio 5s	Group
1	[4]	3.590	119	2013	214.90	523.42	Burchard-Korol group
2a	[41]	3.148	50	2010	171.48	743.95	Lewandowska PUEB group
2b	[41]	3.148	50	2010	171.48	727.88	PUT Poznań group Kłos
3	[58]	8.050	37	2016	147.50	592.34	Univ. of Z. Góra group
4	[43]	1.6	39	2004	130.00	415.66	Foltynowicz PUEB group
5	[45]	3.089	26	2013	116.89	522.51	Kulczycka group
6	[5]	4.900	32	2016	111.00	478.73	Czaplicka-Kolarz et al.
7	[27]	5.715	23	2016	110.15	166.88	Korol et al.
8	[61]	1.08	10	2015	60.80	161.14	Częstochowa LCM Center
9	[63]	0.763	11	2017	38.63	95.87	UTP Bydgoszcz group

Source: own research
[a]Publication number according to the References section

selected. This allowed the ranking of the best publications of Polish authors to be made. Table 11 presents a summary of the best works from individual research groups.

The largest InOrdinatio characterized a work by Burchard-Korol et al. [4], which has been cited 119 times. Second place comes joint publication of authors from PUEB and PUT. The third place is for the group from the University of Zielona Góra. The largest InOrdinatio does not always seem to reflect the actual position of a given group, especially when other publications have smaller InOrdinatio. That is why InOrdinatio was summarized for the five best publications from a given group, resulting in InOrdinatio 5s. It turned out that the leader is Lewandowska group, which accumulated almost 744 InOrdinatio 5s points. The second place with the result of 728 points of InOrdinatio 5s was taken by the team led by Kłos. The third position is occupied by the group from the University of Zielona Góra with 592 InOrdinatio 5s and next (523 InOrdinatio 5s) places are for the Burchard group and Kulczycka group. InOrdinatio was determined using JRC indexes. Perhaps the use of other parametric indexes would affect the ranking results, which will be checked in the future.

It is worth noting that the cooperation of the PUT, PUEB and Polish Academy of Sciences in Cracow groups brings very good scientific and bibliometric results. It is also worth mentioning that Polish scientists are establishing international cooperation, which also brings effects in the form of indexed publications.

One should also mention the numerous monographs on the subject of LCA/LCM by Polish authors, which, however, appeared in Polish. Polish scientists are also co-authors of numerous chapters in monographs. Over 20 doctorates in this field were already defended, and several researchers also obtained postdoctoral degrees. This aspect, however, goes beyond the accepted scope of this study.

References

1. Adamczyk, W. (1999). *Ecobalance – A tool for environmental evaluation of products and manufacturing processes, proceedings of the 12th IGWT symposium* (pp. 670–675). Poznan University of Economic.
2. Adamczyk, J., & Dzikuć, M. (2014). The analysis of suppositions included in the polish energetic policy using the LCA technique—Poland case study. *Renewable and Sustainable Energy Reviews, 39*, 42–50.
3. bazybg.uek.krakow.pl (https://bazybg.uek.krakow.pl/dorobek/welcome/bibliografia/66/0/0/0)
4. Burchart-Korol, D. (2013). Life cycle assessment of steel production in Poland: A case study. *Journal of Cleaner Production, 54*, 235–243.
5. Burchart-Korol, D., Fugiel, A., Czaplicka-Kolarz, K., & Turek, M. (2016). Model of environmental life cycle assessment for coal mining operations. *Science of the Total Environment, 562*, 61–72.
6. Burchart-Korol, D., Korol, J., & Czaplicka-Kolarz, K. (2016). Life cycle assessment of heat production from underground coal gasification. *The International Journal of Life Cycle Assessment, 21*(10), 1391–1403.
7. Burchart-Korol, D., Krawczyk, P., Czaplicka-Kolarz, K., & Smoliński, A. (2016). Eco-efficiency of underground coal gasification (UCG) for electricity production. *Fuel, 173*, 239–246.

8. Danilecki, K., Mrozik, M., & Smurawski, P. (2017). Changes in the environmental profile of a popular passenger car over the last 30 years – Results of a simplified LCA study. *Journal of Cleaner Production, 141*, 208–218.

9. Dylewski, R., & Adamczyk, J. (2014). The comparison of thermal insulation types of plaster with cement plaster. *Journal of Cleaner Production, 83*, 256–262.

10. Dylewski, R., & Adamczyk, J. (2016). Study on ecological cost-effectiveness for the thermal insulation of building external vertical walls in Poland. *Journal of Cleaner Production, 133*, 467–478.

11. Dylewski, R., & Adamczyk, J. (2016). The environmental impacts of thermal insulation of buildings including the categories of damage: A polish case study. *Journal of Cleaner Production, 137*, 878–887.

12. Dzikuć, M., & Adamczyk, J. (2015). The ecological and economic aspects of a low emission limitation: A case study for Poland. *The International Journal of Life Cycle Assessment, 20*(2), 217–225.

13. Flizikowski, J., Piasecka, I., Kruszelnicka, W., Tomporowski, A., & Mroziński, A. (2018). Destruction assessment of wind power plastics blade. *Polimery, 63*, 9.

14. Foltynowicz, Z., & Kłos, Z. (2019). https://www.researchgate.net/publication/338018588_Bibliometric_analysis_of_Polish_LCA's_scientist_performance

15. Foltynowicz, Z., & Lewandowska, A. (2005). Life cycle assessment in Poland – General review. *Forum Ware International, 6*(1), 7–10.

16. Foltynowicz, Z., Kłos, Z., Kurczewski, P., & Lewandowska, A. (2006). *Environmental designing of technical objects as a basis for life cycle management (LCM) – Case Study for Poland, 2nd international conference on quantifies eco-efficiency analysis for sustainability*, 28–30 June 2006 Egmond aan Zee, Netherlands.

17. Fuc, P., Kurczewski, P., Lewandowska, A., Nowak, E., Selech, J., & Ziolkowski, A. (2016). An environmental life cycle assessment of forklift operation: A well-to-wheel analysis. *The International Journal of Life Cycle Assessment, 21*(10), 1438–1451.

18. Fugiel, A., Burchart-Korol, D., Czaplicka-Kolarz, K., & Smolińskic, A. (2017). Environmental impact and damage categories caused by air pollution emissions from mining and quarrying sectors of European countries. *Journal of Cleaner Production, 143*, 159–168.

19. Garfield, E. (2006). The history and meaning of the journal impact factor. *The Journal of the American Medical Association, 295*(1), 90–94.

20. Góralczyk, M., & Kulczycka, J. (2005). LCC application in the polish mining industry. *Management of Environmental Quality: An International Journal, 16*(2), 119–112.

21. Gorazda, K., Tarko, B., Wzorek, Z., Kominko, H., Nowak, A. K., & Kulczycka, J. (2017). Fertilisers production from ashes after sewage sludge combustion–A strategy towards sustainable development. *Environmental Research, 154*, 171–180.

22. Guinée, J. (2016). Chapter 3: Life cycle sustainability assessment: What is it and what are its challenges? In R. Clift & A. Druckman (Eds.), *Taking stock of industrial ecology*. Springer Open. https://doi.org/10.1007/978-3-319-20571-7_3

23. Kłos Z. (1986). Rozważania o celowości wyznaczania środowiskowego kosztu istnienia maszyn i urządzeń. *Zeszyty Naukowe Politechniki Poznańskiej*, seria: Maszyny Robocze i Pojazdy, 1986, no. 26, p. 75–85 (in Polish).

24. Kłos, Z. (1990). *Sozologiczność obiektów technicznych*. Wydawnictwo Politechniki Poznańskiej.

25. Kłos, Z. (1999). LCA in Poland: Background and state-of-art. *The International Journal of Life Cycle Assessment, 7*(5), 249–250.

26. Kłos, Z., & Kurczewski, P. (2009). LCA in Poznań and Poland. Research teams and their achievements. *Scientific Problems of Machines Operation and Maintenance, 2*(158), 85–99. http://t.tribologia.org/plik/spm/spmom-09v44n2_p-085.pdf

27. Korol, J., Burchart-Korol, D., & Pichlak, M. (2016). Expansion of environmental impact assessment for eco-efficiency evaluation of biocomposites for industrial application. *Journal of Cleaner Production, 113*, 144–152.

28. Kowalski, Z., & Kulczycka, J. (2004). Cleaner production as a basic element for the sustainable development strategy. *Polish Journal of Chemical Technology, 6*(4), 35–40.
29. Kowalski, Z., Kulczycka, J., & Wzorek, Z. (2007). Life cycle assessment of different variants of sodium chromate production, Poland. *Journal of Cleaner Production, 15*(1), 28–37.
30. Kulczycka, J. (2009). Life cycle thinking in polish official documents and research. *The International Journal of Life Cycle Assessment, 14*(5), 375–378.
31. Kulczycka, J., & Smol, M. (2016). Environmentally friendly pathways for the evaluation of investment projects using life cycle assessment (LCA) and life cycle cost analysis (LCCA). *Clean Technologies and Environmental Policy, 18*(3), 829–842.
32. Kulczycka, J., Kasprzak, J., Kurczewski, P., Lewandowska, A., Lewicki, R., Witczak, A., & Witczak, J. (2011). The polish Centre for Life Cycle Assessment—The Centre for life cycle assessment in Poland. *The International Journal of Life Cycle Assessment, 5*, 442–444.
33. Kulczycka, J., Lelek, L., Lewandowska, A., & Zarebska, J. (2015). Life cycle assessment of municipal solid waste management--comparison of results using different LCA models. *Polish Journal of Environmental Studies, 24*(1).
34. Kurczewski, P. (2014). Life cycle thinking in small and medium enterprises: The results of research on the implementation of life cycle tools in polish SMEs—Part 1: Background and framework. *The International Journal of Life Cycle Assessment, 19*(3), 593–600.
35. Kurczewski, P., & Lewandowska, A. (2010). ISO 14062 in theory and practice—Ecodesign procedure. Part 2: Practical application. *The International Journal of Life Cycle Assessment, 15*(8), 777–784.
36. Kuzincow, J., & Ganczewski, G. (2015). Life cycle management as a crucial aspect of corporate social responsibility. *Research Papers of the Wroclaw University of Economics / Prace Naukowe Uniwersytetu Ekonomicznego we Wroclawiu, 387*, 91–108.
37. Lelek, L., Kulczycka, J., Lewandowska, A., & Zarebska, J. (2016). Life cycle assessment of energy generation in Poland. *The International Journal of Life Cycle Assessment, 21*(1), 1–14.
38. Lewandowska, A. (2011). Environmental life cycle assessment as a tool for identification and assessment of environmental aspects in environmental management systems (EMS) part 1: Methodology. *The International Journal of Life Cycle Assessment, 16*(2), 178–186.
39. Lewandowska, A., & Foltynowicz, Z. (2004). Comparative LCA analysis of industrial objects part II: Case study for chosen industrial pumps. *The International Journal of Life Cycle Assessment, 9*(3), 180–186.
40. Lewandowska, A., & Foltynowicz, Z. (2004). New direction of development in environmental life cycle assessment. *The Polish Journal of Environmental Studies, 13*(5), 463–466.
41. Lewandowska, A., & Kurczewski, P. (2010). ISO 14062 in theory and practice—Ecodesign procedure. Part 1: Structure and theory. *The International Journal of Life Cycle Assessment, 15*(8), 769–776.
42. Lewandowska, A., & Matuszak-Flejszman, A. (2014). Eco-design as a normative element of environmental management systems—The context of the revised ISO 14001: 2015. *The International Journal of Life Cycle Assessment, 19*(11), 1794–1798.
43. Lewandowska, A., Foltynowicz, Z., & Podleśny, A. (2004). Comparative LCA analysis of industrial objects part I: LCA data quality assurance – Sensitivity analysis and pedigree matrix. *The International Journal of Life Cycle Assessment, 9*(2), 86–89.
44. Lewandowska, A., Wawrzynkiewicz, Z., Noskowiak, A., & Foltynowicz, Z. (2008). Adaptation of ecoinvent database to polish conditions. *The International Journal of Life Cycle Assessment, 13*(4), 319.
45. Lewandowska, A., Kurczewski, P., Kulczycka, J., Joachimiak, K., Matuszak-Flejszman, A., Baumann, H., & Ciroth, A. (2013). LCA as an element in environmental management systems—Comparison of conditions in selected organisations in Poland, Sweden and Germany. *The International Journal of Life Cycle Assessment, 18*(2), 472–480.
46. Lewandowska, A., Noskowiak, A., & Pajchrowski, G. (2013). Comparative life cycle assessment of passive and traditional residential buildings' use with a special focus on energy-related aspects. *Energy and Buildings, 67*, 635–646.

47. Lewandowska, A., Noskowiak, A., Pajchrowski, G., & Zarebska, J. (2015). Between full LCA and energy certification methodology—A comparison of six methodological variants of buildings environmental assessment. *The International Journal of Life Cycle Assessment, 20*(1), 9–22.
48. Marcinkowski, A. (2018). Environmental efficiency of industrial Symbiosis – LCA case study for gypsum exchange. *Multidisciplinary Aspects of Production Engineering – MAPE, 1*(1), 793–800.
49. Marcinkowski, A., & Zych, K. (2017). Environmental performance of kettle production: Product life cycle assessment. *Management Systems in Production Engineering, 25*(4), 255–261.
50. McKiernan, E. C., Schimanski, L. A., Muñoz, N. C., Matthias, L., Niles, M. T., & Alperin, J. P. (2019). Use of the journal impact factor in academic review, promotion, and tenure evaluations. *eLife, 8*, e47338. https://doi.org/10.7554/eLife.47338
51. Muradin, M., & Foltynowicz, Z. (2018). Logistic aspects of the ecological impact indicators of an agricultural biogas plant. *LogForum, 14*(4), 535–547. https://doi.org/10.17270/J. LOG.2018.306
52. Muradin, M., Joachimiak-Lechman, K., & Foltynowicz, Z. (2018). Evaluation of eco-efficiency of two alternative agricultural biogas plants. *Applied Sciences, 8*, 2083. https://doi. org/10.3390/app8112083
53. Nitkiewicz, T., & Starostka-Patyk, M. (2017). Contribution of returned products handling scenarios to life cycle impacts--research case of washing machine. *Environmental Engineering & Management Journal (EEMJ), 16*(4), 1.
54. Pagani, R. N., Kovaleski, J. L., & Resende, L. M. (2015). Methodi Ordinatio: A proposed methodology to select and rank relevant scientific papers encompassing the impact factor, number of citation, and year of publication. *Scientometrics, 105*, 2109–2135. https://doi. org/10.1007/s11192-015-1744-x
55. Pajchrowski, G., Noskowiak, A., Lewandowska, A., & Strykowski, W. (2014). Wood as a building material in the light of environmental assessment of full life cycle of four buildings. *Construction and Building Materials, 52*, 428–436.
56. Pajchrowski, G., Noskowiak, A., Lewandowska, A., & Strykowski, W. (2014). Materials composition or energy characteristic–What is more important in environmental life cycle of buildings? *Building and Environment, 72*, 15–27.
57. Pesonen, H. L., Ekvall, T., Fleischer, G., et al. (2000). Framework for scenario development in LCA. *The International Journal of Life Cycle Assessment, 5*, 21. https://doi.org/10.1007/ BF02978555
58. Piwowar, A., Dzikuć, M., & Adamczyk, J. (2016). Agricultural biogas plants in Poland–selected technological, market and environmental aspects. *Renewable and Sustainable Energy Reviews, 58*, 69–74.
59. Selech, J., Joachimiak-Lechman, K., Klos, Z., Kulczycka, J., & Kurczewski, P. (2014). Life cycle thinking in small and medium enterprises: The results of research on the implementation of life cycle tools in polish SMEs—Part 3: LCC-related aspects. *The International Journal of Life Cycle Assessment, 19*(5), 1119–1128.
60. Śliwińska, A., Burchart-Korol, D., & Smoliński, A. (2017). Environmental life cycle assessment of methanol and electricity co-production system based on coal gasification technology. *Science of the Total Environment, 574*, 1571–1579.
61. Starostka-Patyk, M. (2015). New products design decision making support by SimaPro software on the base of defective products management. *Procedia Computer Science, 65*, 1066–1074.
62. Starostka-Patyk, M., & Nitkiewicz, T. (2014). LCA approach to management of defective products in reverse logistics channel, *2014 International conference on advanced logistics and transport.* https://scholar.google.pl/citations?user=xZJ3Yx8AAAAJ&hl=pl

63. Tomporowski, A., Flizikowski, J., Opielak, M., Kasner, R., & Kruszelnicka, W. (2017). Assessment of energy use and elimination of CO2 emissions in the life cycle of an offshore wind power plant farm. *Polish Maritime Research, 24*(4), 93–101.
64. Tomporowski, A., Piasecka, I., Flizikowski, J., Kasner, R., & Kruszelnicka, W. (2018). Comparison analysis of blade life cycles of land-based and offshore wind power plants. *Polish Maritime Research, 25*(s1), 225–233.
65. Witczak, J., Kasprzak, J., Klos, Z., Kurczewski, P., Lewandowska, A., & Lewicki, R. (2014). Life cycle thinking in small and medium enterprises – The results of research on the implementation of life cycle tools in polish SMEs part 2: LCA related aspects. *The International Journal of Life Cycle Assessment, 19*, 891–900.
66. Żakowska, H. (2004). Wytyczne dotyczące wykonania analizy cyklu życia (LCA) opakowań i ograniczenia tej metody. Guidelines for the performance of the Life Cycle Analysis (LCA) of packages and limitations in this method. *Opakowanie, 11*, 20–24.
67. Żakowska, H. (2014). Metoda LCA w logistyce odzysku odpadów opakowaniowych/LCA method in the logistics of packaging waste recovery. *Logistyka Odzysku, 3*(12), 22–24.

Open Access This chapter is licensed under the terms of the Creative Commons Attribution 4.0 International License (http://creativecommons.org/licenses/by/4.0/), which permits use, sharing, adaptation, distribution and reproduction in any medium or format, as long as you give appropriate credit to the original author(s) and the source, provide a link to the Creative Commons license and indicate if changes were made.

The images or other third party material in this chapter are included in the chapter's Creative Commons license, unless indicated otherwise in a credit line to the material. If material is not included in the chapter's Creative Commons license and your intended use is not permitted by statutory regulation or exceeds the permitted use, you will need to obtain permission directly from the copyright holder.

Index

Printed in the United States
by Baker & Taylor Publisher Services